The Illusion of Linearity

Mathematics Education Library

VOLUME 41

The titles published in this series are listed at the end of this volume.

Dirk De Bock
Wim Van Dooren
Dirk Janssens
Lieven Verschaffel
(Authors)

The Illusion of Linearity

From Analysis to Improvement

Springer

Dirk De Bock
Centre for Instructional Psychology and Technology
K.U. Leuven, Belgium and EHSAL—European University College Brussels
Belgium

Wim Van Dooren
Centre for Instructional Psychology and Technology
K.U. Leuven
Belgium

Dirk Janssens
Department of Mathematics
K.U.Leuven
Belgium

Lieven Verschaffel
Centre for Instructional Psychology and Technology
K.U. Leuven
Belgium

ISBN -13: 978-0-387-71082-2 e-ISBN-13: 978-0-387-71164-5

Library of Congress Control Number: 2007922630

Printed on acid-free paper.

9 8 7 6 5 4 3 2 1

springer.com

Contents

Contributing Authors

Dirk De Bock
Centre for Instructional Psychology and Technology, K.U.Leuven, Belgium and EHSAL – European University College Brussels, Belgium

Wim Van Dooren
Centre for Instructional Psychology and Technology, K.U.Leuven, Belgium

Dirk Janssens
Department of Mathematics, K.U.Leuven, Belgium

Lieven Verschaffel
Centre for Instructional Psychology and Technology, K.U.Leuven, Belgium

Foreword

In a cubic centimeter, there are 1 000 cubic millimeters, in a cubic decimeter 1 000 000, in a cubic meter 1 000 000 000, and so forth. Why on earth is it usually so difficult to teach and to learn such simple facts, and many others of a similar vein? Of course, some questions of this type are more intricate. There is no easy computation showing that a giant ten times as high as a given dwarf weighs about one thousand more. So the nature of the problem is a crucial factor, and the authors of this study are fully aware of that.

This book deals with the illusion of linearity, mainly in the context of enlargement and reduction of figures and solids. An elementary example is when somebody believes that multiplying the side of a square by 2 implies that its area is also multiplied by 2. The book approaches also, to some extent, the context of probabilities. The authors rely essentially on two methods of investigation, namely experiments involving an experimental and a control group of students, and individual interviews on the other. A number of important variables are scrutinized, the most important being:

- drawings made by students themselves versus ready-made drawings, using squared paper or not
- *direct* versus *indirect* measures
- problems stated in the missing-value format, or in the comparison format
- awakening students' consciousness by a preliminary significant question
- degree of authenticity of the situation.

Two of these variables deserve some further explanation.

Direct versus indirect measures. Expressing an area in square meters is an example of a *direct measure*, while relating the area of a surface to the amount of paint required to cover it is an example of an *indirect measure.*

Missing-value or comparison format. As an example, one side of a polygon measures 2 dm and its area is 6 dm^2. What happens to the area in an enlargement operation where the 2 dm become 8 dm? This is a missing-value format. And if the 2 dm were increased by a factor of 4? This is a comparison format. In short, the givens are three measures in the former case, and two measures and a ratio of measures in the latter.

An impressive result of the study is how deep-rooted the illusion of linearity is and how strongly it resists many variations of the teaching and learning parameters. The principal circumstance in which the illusion is substantially weakened is when the students are asked, instead of drawing or computing, to physically cover the enlarged surface (of the problem) by an appropriate paving. And even in that case, the improvement of the students' awareness does not substantially withstand returning to more academically stated questions.

The authors also tried a series of ten classroom one-hour sessions inspired by the principles of realistic mathematics education: meaningful and attractive problems, small group work, whole-class discussions and, as far as mathematical matters are concerned, a variety of representations and symbols (drawings, tables of functions, graphs, formulas). Even such a more concentrated and well-oriented pedagogical action did not yield entirely satisfactory results: "many students did not develop a deeper understanding of (non)-proportionality."

For such persistence of the illusion of linearity, three main causes are identified. One of them pertains to the way proportionality is often taught, namely when some parts of the curriculum pay an almost exclusive attention to proportionality as compared to non-linear relations, when there is an overuse of missing-value problems and an overemphasis on routine solving processes as compared to meaningful analysis of situations. Indeed, proportionality is more than a four-term relation. There are the classical rule of three, tables of proportionality showing more than four terms, straight-line graphs passing through the origin, the constant slope of such graphs identified with the coefficient of proportionality, etc. Of course, these features are better understood when contrasted with non-proportional (non-linear) relations. If the teaching ignores these meaningful facets of linearity and remains confined to the narrow domain of four-term missing-value questions, and if it does not contrast proportionality with non-proportionality, then the students are likely to remain like short-sighted prisoners in an obscure intellectual cell. As was so convincingly explained by Wertheimer (1945), a perspicacious problem solver in a given domain is one who knows the landscape familiarly, i.e. not only its various parts, but the ways to circulate amongst them. Perceiving the very structure as a whole is crucial.

Further, if linearity and non-linearity ought to be regularly confronted, doesn't it mean that the real notion at stake is the one of function with its various modalities? As Klein (1939) wrote a long time ago (the quotation remains surprisingly timely after such a long period):

> We, who are readily called reformers, want to place the concept of function at the centre of teaching. For it is that mathematical concept of the last 200 years which, wherever mathematical thought is needed, plays a central role.

Another cause of the persistent illusion of proportionality can be found, according to the authors, in some shortcomings of the general geometrical knowledge of the students. However, this second cause is akin to the first one. One refers to the teaching of proportionality, the other to an unsuccessful teaching of geometry in general. But what does it mean that, when solving proportionality questions, the students show some gaps in their general geometrical knowledge? It means that their understanding of proportionality lacks some structural links with significant adjoining geometrical questions. Generalizing this comment, one might say that mathematics is not a juxtaposition of items, it is an integrated culture. What is at stake is the mobility of mind. The authors are aware of that. As a remedy, they propose to displace the emphasis from computing correct numerical answers to building appropriate mathematical models. But what is the substance of such models if not those mathematical notions and properties that faithfully express the structure of the situation on hand?

Let us now leave aside the deficiencies of the teaching system. A third cause of the illusion of linearity is of a more intrinsic nature. It relates to the intuitiveness and simplicity of the linear relation. This deserves some comments. Let us assume that a student received a fully appropriate instruction on proportionality and non-proportionality. She or he might still be seduced by the charms of proportionality. This is an effect of what might be called the inertia of concepts. Proportionality is similar, to some extent, to a paradigm in the sense of Kuhn (1962). When you have an intellectual instrument at your disposal, if this instrument properly solved a lot of previous problems, if it appears simpler and more elegant than others, then you stick to it until further notice. What happens here pertains simultaneously to the pleasant and simple nature of the knowledge and the indolent nature of the human mind. The inertia of the concepts is also illustrated by a striking finding of the study. In fact, when students have been duly trained, on a number of examples, to identify the non-linear situations, they show a tendency to overuse a non-linear model. Changing ones mind is not so easy, but once this is done... Le mieux est l'ennemi du bien[1].

Now, what could be done to avoid such seductions, if not developing the habit of doubting, a critical mind, a constant circumspection in front of any problem, the habit of checking everything? To conclude, may I express how I appreciate the honesty of this study. It brings us a most careful survey of a number of real difficulties more than a wealth of solutions. All the more so that one of the findings is that even ten classroom sessions are not enough to bring a persistent change. No panacea is proposed. These questions have to be considered in a long run perspective. In the mean time, some doubts will remain. But after all, as Dante wrote in the Inferno,

> Che non men che saper, dubbiar m'aggrada,

which means

> As well as knowing, doubting is praiseworthy.

Nicolas Rouche
Professor Emeritus of the Catholic University of Louvain, Belgium

[1] The best is the enemy of the good.

Acknowledgments

In this monograph we report and reflect upon a series of studies that have been done over the past years on the 'illusion of linearity'. Although this book is the product of several years of intensive collaboration between the four authors, many other people had played a part in shaping the overall research program, in designing the experiments, in analysing and interpreting the outcomes, and – last but not least – in preparing the manuscript.

Special thanks go to Nicolas Rouche and Brian Greer for their comments and advice given in various stages of the research cycle reported in this book. More particularly, we thank Professor Rouche for his clear and inspiring preface to this book and Professor Greer for his generosity – both as a native speaker and an expert in the domain of the psychology of mathematics education – in giving the whole manuscript a final check.

We owe thanks to An Hessels, who worked for several years as a research assistant on the project, and who had a significant contribution in several studies reported in the book, as well as to Karen Claes, Elke De Bolle, Fien Depaepe, and Rebecca Rommelaere, who were as Master students involved in a particular study.

We also want to thank the series editor Alan Bishop and the members of the Editorial Board for endorsing this publication project and for their valuable comments on the concept and on the first draft of the book, the anonymous reviewers, and the publisher who worked hard to move this book quickly through to publication.

Furthermore, we are grateful to the Research Fund of the University of Leuven (Belgium) for the financial support to this project by grant OT 2000/10 "The illusion of linearity in secondary school students: From analysis to improvement" and, afterwards, by grant GOA 2006/01 "Developing adaptive expertise in mathematics education".

Finally, special thanks are due to Karine Dens for the competence, patience, and care with which she has helped us in preparing the text for publication.

Dirk De Bock
Wim Van Dooren
Dirk Janssens
Lieven Verschaffel

December 2006

Chapter 1

A WIDESPREAD PHENOMENON

According to legend, the citizens of Athens consulted the oracle of Apollo at Delos in 430 BC, to learn how to defeat a plague which was ravaging their lands. The oracle responded that to stop the plague, they must double the size of their altar. The Athenians dutifully doubled each side of the altar, but the plague increased (Smith, 1923).

Aristotle believed that the speed an object falls at is proportional to its weight. So if you have a ball weighing 100 g, and one weighing 1 kg, the heavier one will fall ten times as fast, when released from the same height. It took many centuries before he was proven wrong by Galileo Galilei (Galilei, 1638).

Cardano, the Italian Renaissance scholar, turned to gambling to boost his financial situation, and because of his understanding of probability, he won more than he lost. Nevertheless, he reasoned that one has to roll two dice 18 times to have a probability of 50% to get a double-ace at least once (Székely, 1986).

1. INTRODUCTION

Linear or direct proportional relationships receive a lot of attention in contemporary mathematics education, both at the elementary and secondary level. The reason lies in the fact that linear relationships are the underlying model for approaching numerous practical and theoretical problem situations within mathematics and science. In their search for a *fil conducteur* (a general guiding principle) throughout mathematics education, the Centre de

Recherche sur l'Enseignement des Mathématiques in Nivelles, Belgium (2002, p. 2) concluded that

> l'idée de linéarité, qui apparaît modestement à l'école maternelle, se construit par généralisations successives tout au long de la scolarité[2].

Linearity passes through the entire mathematical building, from the idea of measuring magnitudes, the concept of ratios, and the application of the 'rule of three' in primary school to linear algebra and the use of linear models in calculus and statistics in secondary school, and to the abstraction in a vector space sense in higher education.

As proportionality is, on the one hand, one of the most elementary higher order understandings and, on the other hand, one of the most advanced elementary understandings, the skills of proportional reasoning play a 'watershed role' (Lesh, Post, & Behr, 1988) in students' mathematical development. Not surprisingly, the mathematics education research literature abounds with studies on the development of students' multiplicative and proportional reasoning, on problems that can occur in this development, and on didactical approaches to teach proportional reasoning (e.g, Behr, Harel, Post, & Lesh, 1992; Karplus, Pulos, & Stage, 1983a; Noelting, 1980a, 1980b; Streefland, 1984; Tourniaire & Pulos, 1985).

Students' growing proportional reasoning expertise and their increasing familiarity with linear models may, however, have a serious drawback. Freudenthal (1983, p. 267) has warned that

> linearity is such a suggestive property of relations that one readily yields to the seduction to deal with each numerical relation as though it were linear.

The tendency of students to apply properties of linear relations 'anywhere' – thus also in situations where this is inadequate – has also been called the 'illusion of linearity', the 'linearity trap', the 'linear obstacle', etc. It has been mentioned and commented upon in numerous practical documents and research reports, and reported in students of various ages and in diverse mathematical domains. But until recent years there were no systematic investigations of this phenomenon. Consequently, little was known about the mathematical, psychological, and didactical factors that are responsible for the occurrence and persistence of this phenomenon and about how it can be countered by appropriate instruction.

[2] The idea of linearity, appearing modestly in Kindergarten, constructs itself by successive generalizations throughout the school career.

In the last 5-10 years, substantial research efforts have been undertaken to fill this gap in our knowledge about students' overuse of linearity. The overall goal of the present monograph is to present, and reflect upon, the current state-of-the-art of the research on this phenomenon. This research has not only shown the impact of students' overuse of linearity in various mathematical problem-solving contexts, but has also deepened our understanding of the psychological and educational roots of this phenomenon (in the domain of geometry but also in other mathematical domains). In this opening chapter, we first investigate the concept of linearity itself in terms of its properties and representations. Second, based on the available literature, we summarize and comment on several cases of unwarranted applications of these linear properties and representations in diverse mathematical domains. Most often, the reports of these unwarranted applications of linearity reported in the literature have an exemplary and anecdotal character, but in a few cases they have been studied more profoundly by cognitive psychologists or mathematics educators. In the last section of this first chapter, we focus on the overuse of linearity in the domain of geometry and we sketch the research cycle that lies at the basis of this monograph.

2. PROPERTIES AND REPRESENTATIONS OF LINEARITY: A CONCEPTUAL ANALYSIS

Throughout this monograph, we will reserve the terms 'linear' and '(direct) proportional' to refer to functions of the form '$f(x) = ax$' (with $a \neq 0$) and we will use these terms – as well as their derivatives 'linearity' and '(direct) proportionality' – as synonyms. The graphical representation of this type of function is well-known: a straight line passing through the origin. But there are other well-known and frequently used properties and representations of linear functions. We will elaborate on each of them here, since the different representations may emphasize (and de-emphasize) particular properties of the linear system (see e.g., Lesh & Doerr, 2003) and each of these properties and representations can also be applied by students in situations where it is not applicable.

There is a close link between (direct) proportionality or linearity and the concepts of ratio and proportion (Centre de Recherche sur l'Enseignement des Mathématiques, 2002). A ratio refers to a fractional relation between two measures (a/b), for example: "This lemonade was made with 10 oranges for each litre of water" or "the population grows by 1000 people each year". A proportion refers to the equality of two ratios $a/b = c/d$ (and thus to a relation between four measures), such as: "In one urn there are 20 white balls and 40 black balls. The other urn has 50 white and 100 black balls. The chance for drawing a white ball is equal for both urns since 20/40 = 50/100." The concept of linearity or proportionality (or a linear or proportional *relation* between quantities) refers to the equality of a multitude of equal ratios: $a/b = c/d = e/f = \ldots$. For example, "a can of 0.75 litre of paint is enough for painting 6 m^2, so I need 1.50 litre for 12 m^2, 1.75 litre for 14 m^2, etc.". In the functional sense, it refers to a constant ratio between the independent and the dependent variable, which means that each two ratios that are 'picked out' will yield a proportion: e.g., 14/1.75 = 6/0.75.

Two essential properties of linear functions in the above-mentioned sense are that $f(x + y) = f(x) + f(y)$ and $f(kx) = k\,f(x)$. For covering 18 m^2 (which is 6 m^2 + 12 m^2), I need 2.25 litre (0.75 litre + 1.50 litre) of paint, and for 28 m^2 (2 × 14 m^2) I need 3.50 litre (2 × 1.75 litre) of paint. Usually, in more advanced approaches to linear algebra, linear relations are even *defined* in terms of these two properties.

The multiplicative property above, namely $f(kx) = k\,f(x)$, refers to the equality of internal ratios and is often represented in the '*k* times *a*–*k* times *b*' rule, which is generally put in the forefront in the classroom teaching when dealing with relatively simple problem situations, for example "To cover 3 times as much area, I need 3 times as much paint, and with 10 times less paint, I can cover 10 times less area".[3] Often, '*k* times *a*–*k* times *b*' reasoning is taught by means of 'arrow schemes' representations. Note that when the proportionality factor is a (relatively small) integer, especially younger students may tend to focus on the additive property (e.g., "for 6 + 6 + 6 m^2, I need 0.75 + 0.75 + 0.75 litre of paint") because it corresponds more to their prior learning experiences. That type of reasoning itself essentially relies on the linear properties of the problem situation. In this respect, Kaput and West (1994) speak of 'competent, but informal proportional reasoning'.

Vergnaud (1983) developed the notion 'measure space' to assist in clarifying the multiplicative relations that exist in proportional situations. A measure space is a physical magnitude (such as area or amount of paint) for which quantities – expressed with a certain unit of measure – exist (e.g., 6 m^2, 0.75 litres). Proportionality can be seen as a multiplicative relationship between the quantities within two different measure spaces. In the above example, the quantities between the two measure spaces are related by multiplication, which refers to the external ratios (multiplying the litres of paint by 8 gives the number of square metres that can be covered with it,

[3] A corresponding rule is also frequently applied in the case of inverse proportional relations being phrased as a '*k* times more *a*, *k* times less *b*': "if I walk 3 times as fast, it will take me 3 times less time to reach my destination".

hence the relation between paint and area is of the form $f(x) = ax$), and a multiplicative relation also exists between the elements within each measure space (e.g., "with 3 times as much paint, I can cover 3 times as much area", which refers to the $f(kx) = k\,f(x)$ property).

Often, functions of the form '$f(x) = ax + b$' are also named 'linear'. Such functions are graphically represented by a straight line that does not pass through the origin except in the special case $b = 0$. In this book, we make the choice to use the term 'linear' in its narrower sense, i.e. for functions of the form '$f(x) = ax$' , and to avoid confusion, we will label functions of the form '$f(x) = ax + b$' with $b \neq 0$ as 'affine' functions. The reason that we use a narrow interpretation is that students may also tend to rely on properties or representations of linear functions (e.g. the $f(x + y) = f(x) + f(y)$ and $f(kx) = k\,f(x)$ properties or the representation by a straight line through the origin) when they are actually dealing with affine situations, for which these properties do not hold.

To summarise, this book considers functions of the form $f(x) = ax$ as linear (or proportional) relations. There are several interrelated characteristics and representations of linear functions that students may apply while dealing with this type of functions:

- Linear functions are represented by a straight line through the origin.
- Any two ratios within this linear function yield a proportion, so that a 'typical' proportional missing-value word problem (such as: "12 eggs cost 2 euro. What is the price of 36 eggs?"), can be solved by calculating the missing value x in a proportion (here $12/2 = 36/x$).
- For each linear function, it holds that $f(x + y) = f(x) + f(y)$ and $f(kx) = k\,f(x)$.
- The multiplicative property is often put to the forefront by a 'k times a–k times b' rule.
- In situations with an integral and relatively small proportionality factor, the additive property (e.g., '$a + a$, so $b + b$' or '$a + a + a$ so $b + b + b$') is often applied instead of the multiplicative property ('k times a–k times b').

As will be amply shown in the next section, each of these characteristics and representations can also be applied (either separately or in combination) by students in situations where it is not adequate.

3. MANIFESTATIONS OF OVER-RELIANCE ON LINEARITY

At several places, the practically-oriented and scientifically-oriented literature on mathematics education (and occasionally also the literature on science education) mentions students of different ages tending to rely on one or several of the above-mentioned linearity characteristics and/or representations in non-linear situations belonging to a variety of mathematical subdomains. So far, there was no attempt to collect these exemplary manifestations and to search for similarities and dissimilarities among them. Such a synthesis is the goal of the current section. It is, however, impossible to give an exhaustive overview of all manifestations of over-relying on linearity in mathematical reasoning that we have encountered in our survey of the literature. Rather, we will show how diverse and universal these manifestations are, and make some conceptual distinctions, keeping in mind the different characteristics and representations of linear relations described in the previous section.

We want to stress that while some manifestations of students' overuse of linearity have been the subject of empirical research, most of them were only reported as incidental findings in publications that actually had another focus. This implies that in some cases we can give a rather detailed, research-based description and interpretation of the committed error whereas in other cases we are unable to do so. The cited examples, moreover, originate from studies with substantially different theoretical backgrounds.

3.1 Improper applications of linearity while solving arithmetic word problems

Typically, from 4th grade on, students are frequently confronted with proportionality problems, most often stated in a missing-value format. Often, it is claimed that such word problems can act as a substitute for the everyday-life situations in which students will need these mathematical skills (Verschaffel, Greer, & De Corte, 2000). Several studies have shown that students have a tendency to use proportional solution methods also for solving arithmetic word problems for which they are not appropriate. Roughly speaking, this phenomenon has been observed in three types of studies: (1) studies about students' lack of sense-making in solving mathematics word problems, (2) studies on students' reasoning in ratio and proportion tasks, and (3) studies explicitly focusing on students' over-reliance on linearity.

The first category of studies has mainly been aimed at unraveling students' 'suspension of sense-making' in school mathematics (e.g., Greer, 1993; Verschaffel, De Corte, & Lasure, 1994; for an overview see Verschaffel et al., 2000). In these studies, upper elementary students were confronted, among other problem situations, with so-called 'pseudoproportionality' problems (e.g., "John's best time to run 100 meters is 17 seconds. How long will it take him to run 1 kilometer?", "A shop sells

312 Christmas Cards in December. About how many do you think it will sell altogether in January, February and March?"). Only very few students appeared to show awareness that direct proportionality will not give the exact answer but only – in the best case – an approximate answer (for the above-mentioned runner problem, percentages of students showing such awareness ranged from 0% to 7% in a range of replications in many countries) (Verschaffel et al., 2000). Typically, this kind of problem elicited many answers based on a *k* times *a*–*k* times *b* reasoning (e.g., for the first problem: "John needs 10 times as much time to run 10 times as far") or on its informal building-up variant (e.g., for the second problem: "The shop will sell 312 + 312 + 312 = 936 cards"). Rather than taking into account their common-sense knowledge and realistic considerations about the situation described in the problem, the students simply play the 'game of school word problems', in which the players are assumed not to attend too much to the realities of the problem situation but merely to identify the arithmetic operation(s) with the given numbers yielding the supposedly correct answer (Verschaffel et al., 2000; see also Nesher, 1996; Reusser & Stebler, 1997a; Wyndhamn & Säljö, 1997). A problem with the interpretation of students' proportional answers in these studies is, however, that most of these problems were 'unsolvable' (i.e. there is no precise logico-mathematical relation between the givens in these items, so an exact answer to the problem cannot be given), and therefore 'unusual' or even 'tricky' because students do not expect 'unsolvable' problems in a testing context. So, it is unsure that students in these studies really 'believed' that there was a proportional relation at stake. Taking into account the rules of the game of school word problems – or, as Brousseau (1997) would state it, due to the working of the 'didactical contract' – students solving these problems may have assumed that these problems had an exact, numerical answer and that they should do some calculations with the numbers given in the problem to provide that answer. There is ample empirical evidence for this phenomenon (for an extensive overview, see Verschaffel et al., 2000).

A second category of studies in which students were reported to unwarrantedly apply linearity in solving word problems, comprises studies that explicitly deal with the teaching and learning of proportional reasoning. During the last twenty years, a lot of research has been conducted in this area (for reviews, see, e.g., Behr, et al., 1992; Litwiller & Bright, 2002; Tourniaire & Pulos, 1985), including research on students' errors and primitive strategies, like the well-known errors due to additive reasoning (e.g., "3 oranges to 5 parts of water taste the same as 7 oranges to 9 parts of water", see Hart, 1981; Karplus, Pulos, & Stage, 1983b; Lin, 1991). A few authors working in this domain have mentioned the phenomenon of unwarranted proportional reasoning and warned about it. Schwartz and Moore (1998, p. 475) stated that

> when proportions are placed in an empirical context, people do not only need to consider at least four distinct quantities and their potential relationships, they also need to decide which quantitative relationships are relevant.

The example they gave relates to mixing 1 oz. of orange concentrate and 2 oz. of water, compared to mixing 2 oz. of orange concentrate and 4 oz. of water. If the question is which mixture will taste stronger, the ratios should indeed be compared, but if the question is which mixture will make more, a ratio comparison is of course inappropriate. Along the same lines, Cramer, Post, and Currier (1993, p. 160) argued that "we cannot define a proportional reasoner simply as one who knows how to set up and solve a proportion" and diagnosed that textbooks do not sufficiently emphasize the ability to discriminate linear and non-linear situations. They confronted 33 preservice elementary teachers with the following additive problem: “Sue and Julie were running equally fast around a track. Sue started first. When she had run 9 laps, Julie had run 3 laps. When Julie had completed 15 laps, how many laps had Sue run?” and observed that thirty-two of the pre-service elementary teachers responded to this problem by setting up and solving a proportion: $9/3 = x/15$; $3x = 135$; $x = 45$. In contrast with the ‘pseudoproportionality’ problems that were used in the previously-mentioned studies on students’ suspension of sense-making in a mathematics classroom, the problem used by Cramer et al. (1993) is clearly ‘solvable’; moreover, these pre-service teachers possessed all necessary mathematical tools to solve it. In our interpretation, what lured them into the proportionality trap was the presentation of this additive situation in a missing-value format. Although there is as such nothing wrong with this formulation, it strongly cues proportional schemes and procedures because most proportional reasoning tasks students encounter in their school careers are stated in a missing-value format (whereas non-proportional problems are rarely, if ever, stated in a missing-value format).

We are only aware of one study belonging to the third category, i.e., investigations that explicitly address the overuse of linearity in students’ solving of arithmetic word problems. Van Dooren, De Bock, Hessels, Janssens, and Verschaffel (2005) investigated when the tendency to over-rely on linearity for solving such problems originated and how it developed with age in relation to students’ learning experiences and their emerging proportional reasoning skills.

They confronted a large group of students from Grade 2 to 8 with a test consisting of 8 missing-value word problems: 2 proportional ones (for which a proportional solution was correct) and 6 non-proportional ones belonging to three different categories: 2 additive, 2 affine and 2 constant problems. Here, we give an example from each category of non-proportional problems:

- Additive problem: "Ellen and Kim are running around a track. They run equally fast but Ellen started later. When Ellen has run 5 laps, Kim has run 15 laps. When Ellen has run 30 laps, how many has Kim run?" (correct answer: 40 laps, proportional answer: 90 laps)
- Affine problem: "The locomotive of a train is 12 m long. If there are 4 carriages connected to the locomotive, the train is 52 m long. How long is the train if there are 8 carriages connected to the locomotive?" (correct answer: 92 m, proportional answer: 104 m)
- Constant problem: "Mama put 3 towels on the clothesline. After 12 hours they were dry. Grandma put 6 towels on the clothesline. How long did it take them to get dry?" (correct answer: 12 hours, proportional answer: 24 hours)

The results showed that, although 2nd-graders already showed some emerging proportional reasoning skills, most progress in correctly solving the proportional word problems was made between 3rd and 6th grade. With respect to the non-proportional problems, more than one third of all answers involved the erroneous application of the proportional model. Not surprisingly, the tendency to over-rely on proportionality developed in parallel with the ability to solve proportional word problems: It was noticeable already in 2nd grade, but increased considerably in the next years, with a peak in 5th grade, where more than half of the answers to the non-proportional problems were proportional errors. After this peak, the number of proportional errors gradually decreased, but they did not disappear completely. In 8th grade still more than one fifth of the answers reflected a misapplication of proportionality.

There were some remarkable differences according to the mathematical model underlying the non-proportional problems. One would expect that the word problems with a 'constant' model (like the 'clothesline' problem mentioned above) would be the easiest problems in the test (since there was no need for calculations), but they got the highest rate of proportional errors (up to 80% in 5th grade). For some word problems (like the additive 'runners' problem), the percentage of correct answers even *de*creased by 30% from Grade 3 to 5, whereas the percentage of erroneous proportional answers increased accordingly.

This study by Van Dooren et al. (2005) showed that primary school pupils strongly tend to apply proportional solution strategies when confronted with non-proportional missing-value word problems in the domain of elementary arithmetic. The tendency already emerged in the 2nd grade, but increased considerably up to 5th grade. By then, pupils had received intensive training in solving proportionality problems. Although the size of the increase differed between the distinct problem categories, this trend was general.

3.2 Over-reliance on linearity in graphical environments

Students' excessive adherence to linearity has also been observed in graph sketching activities. For example, Leinhardt, Zaslavsky, and Stein (1990) categorized students' errors in sketching of graphs and labelled one of their error categories as 'linearity'. These authors also mentioned several studies showing that students of different ages have a strong tendency to produce a linear pattern through the origin when asked to graph non-linear relations, for example the growth of the height of a person from birth to the age of 30. Markovits, Eylon, and Bruckheimer (1986) discovered that 14–15-year old students who were asked to generate examples of functions, or to draw graphs of functions passing through given points, stuck to linear functional thinking. Recently, another study has shown that also preservice elementary school teachers had the misconception that the notion of function just refers to a linear relation (Evangelidou, Spyrou, Elia, & Gagatsis, 2004).

Besides graph sketching, students also often stick too strongly to linearity when judging from a graphical representation whether there is a relation between two variables. For example, in a teaching experiment with 11th-graders, Van Deyck (2001) offered students a scatter plot in which the pattern of the dots showed a parabola. Although these students already had encountered several non-linear models in their mathematics curriculum (including quadratic models), they typically argued that there was no relation between the two represented variables, and they saw a further argument for the absence of a relationship in the Pearson correlation coefficient in the corresponding data set being nearly zero. Apparently, when looking at the graph, the students only searched for a straight-line pattern, and they were moreover not aware that a Pearson correlation coefficient can indicate the presence of linear relations only.

Although, from a mathematical point of view, linearity and straight-line graphs are closely related, it remains unclear whether students always make this link, and really conceive the relation that they represent as a linear one. The straight line may merely act for students as a prototypical representation for any relation between magnitudes, just like an arrow scheme representation acts as a panacea for solving word problems with a missing-value structure. We are not aware of any research evidence showing that students' misuse of linearity in numerically or verbally described situations (e.g., in word problems), on the one hand, and in graphical environments, on the other hand, have the same cognitive roots, but both do have in common that a particular representation of linear relations is used outside its applicability range.

3.3 Improper application of linearity in probabilistic situations

As argued by many authors (e.g., Konold, 1989; Shaughnessy, 1992), the domain of probabilistic reasoning is very sensitive to the presence of phenomena that they call 'misconceptions', 'fallacies', 'biases', etc. Several of those errors can be interpreted as improper applications of proportionality (Van Dooren, De Bock, Depaepe, Janssens, & Verschaffel, 2003b). A typical example is students' belief that the probability of at least one success in a game of chance is proportional to the number of trials (e.g. the probability of 'at least a six' in 1 roll with a fair die equals 1/6, the probability in 2 rolls is $2 \times 1/6$, in 3 rolls it is $3 \times 1/6$ etc.). This problem is related to the error made by Cardano explained at the beginning of the chapter. Correctly assuming that the probability of getting a double-ace when rolling two dice is 1/36, he reasoned that one has to roll the dice 18 times to have a probability of 50% of getting a double-ace at least once (Székely, 1986). Another historical example is the 'problem of the dice'. Chevalier de Méré believed that it should be equally advantageous to bet on the event 'at least one double-six in 24 rolls of 2 dice' than it is to bet on 'at least one six in 4 rolls of a fair die', reasoning that from 4 to 24 rolls, the number of opportunities to get a success increases by factor 6, but, correspondingly, the likelihood of getting a success is divided by 6, so that the effects would cancel out. Freudenthal (1973, p. 585) sharply commented on this faulty reasoning:

> He [de Méré] applied the mathematics he knew, the kind of mathematics which in my childhood was called the rule of three… Maybe he would have performed better if he had never learned mathematics at all!

Actually, it is not surprising that linearity errors occur in the domain of probability. The history of probabilistic thinking is full of mathematicians making errors. Although chance situations obey certain mathematical principles, immediate and conclusive feedback on the correctness of one's assumptions is often not available. Our claim that errors in probabilistic reasoning are often overgeneralisations of linear properties is supported by the fact that intuitively, the notions of 'chance' and 'proportion' are so closely related (Fischbein & Gazit, 1984; Truran, 1994).

> Since comparing probabilities entails the comparison of two fractions, proportional reasoning is considered to be a basic tool of probabilistic reasoning (Lamprianou & Lamprianou, 2002, p. 273).

Recently, the presence of the overuse of linear properties in probabilistic reasoning was investigated by Van Dooren et al. (2003b). Based on a review of the literature on probabilistic reasoning, these authors made an inventory of errors that can be conceptually related to the application of linearity. A typical example is the well-known 'birthday paradox' (see, e.g., Hawkins & Kapadia, 1984; Shaughnessy, 1992): When estimating the probability that at least two students in a class of 30 have the same birthday, many students consider the ratio between the number of 'available' birthdays in the classroom (30) relative to the number of possible birthdays (365 in a regular year), leading to a probability judgement of 30/365 = 0.08, which is a large underestimation of the real probability. They implicitly assume a linear relation between the number of students in the classroom and the probability of a double birthday, and they indeed:

> have the intuition that they would need 180 students or more to have an even chance of getting a match of birthdays (Shaughnessy, 1992, p. 480)

because 180/365 ≈ ½. Within their inventory of linearity-related probabilistic errors, Van Dooren et al. (2003b) identified several erroneous reasonings that could be explained by an improper assumption of a linear relation between the variables determining a binomial chance situation, such as the above-mentioned 'problem of the dice' of de Méré or the well-known 'hospitals problem' from Kahneman and Tversky (1972; see also Fischbein & Schnarch, 1997).

The study of Van Dooren et al. (2003b) also provided empirical evidence that the over-reliance on linearity is indeed the explanation of students' errors in these situations. They confronted large groups of 10th and 12th grade students with a test consisting of several problems situated in the context of rolling fair dice. The 12th-graders had received an introductory course in probability (including the binomial probability model), while the 10th-graders had not. In the test, participants had to make both qualitative and quantitative comparisons of the probabilities of two events. A qualitative comparison means that the participants had to indicate whether a first event had a higher, lower or equal probability as a second event (e.g. "Is the chance of getting a three at least 2 times when rolling a die 4 times larger than, smaller than or equal to the chance of a three at least 2 times when rolling the die 5 times?"). For the quantitative comparisons, the researchers proposed quantitative comparisons of two events and the participants had to judge the correctness of these quantifications (e.g. "The chance of getting a six at least 2 times when rolling a die 12 times is 3 times as large as the chance of getting a six at least 2 times when rolling a die 4 times. Is this true or not true?"). Because the quantifications in these items were proportional, the students always had to judge them as incorrect. For all problems in the test, students were asked to motivate their answer.

The large majority of the students in both age groups performed very well on the qualitative comparisons. In about 90% of the cases, the correct alternative was chosen, indicating that even before formal instruction in probability, students have a good qualitative understanding of how the probability in a situation changes when an aspect of this situation changes. The 12th-graders, who had already met the binomial probability distribution in their curriculum, performed only slightly better on these qualitative comparison problems than the 10th-graders (89.5% versus 88.2% correct answers). The high performance on the qualitative comparisons was in sharp contrast with the low score on the quantitative comparison problems. For this last category, students most frequently chose the incorrect alternative, which assumed proportional relationships. Apparently, the vast majority of the students agreed with an incorrect proportional quantification of their correct quantitative insights. The 12th-graders performed slightly better on the quantitative problems (on the average, 25.8% correct answers) than 10th-graders (18.2% correct answers) but, as expected, the tendency towards linear reasoning modelling was still strongly present in these older students.

A qualitative analysis of the written notes and explanations accompanying the incorrect answers on the quantitative comparison problems revealed that more than 80% of the incorrect answers on these problems could be clearly identified as resulting from students' over-reliance on the linear model. Examples of such statements were:

> In the first case, you can try three times more to obtain the same result (two sixes), so it is evident that you have three times more chance of winning.

> The chance of getting 2 sixes in 12 trials is a lot bigger than getting 2 sixes in 4 trials. And 12 is three times larger than 4, so the statement is true.

In sum, the study by Van Dooren et al. (2003b) provided empirical evidence that secondary school students have a good qualitative understanding of probabilistic situations and are able to compare two such situations that differ in one variable (e.g., in the context of rolling dice, they understand that the chance of having at least one six increases with the number of trials). This understanding is even present in students without formal instruction in probability. At the same time, however, most students have a strong tendency to incorrectly quantify their correct qualitative insights as linear relationships between the variables in a binomial chance situation. This tendency was strongly present in all students in the research group, even in those who had met the binomial probability distribution in their mathematics curriculum.

3.4 Number patterns, algebra, and calculus

Some investigators of number patterns and algebraic generalization report that students over-rely on certain properties of linear models. Stacey (1989) studied 9–13-year old students' modeling of affine patterns of the form $f(n) = an + b$ (with $b \neq 0$) such as the following 'ladders' problem (Figure 1-1).

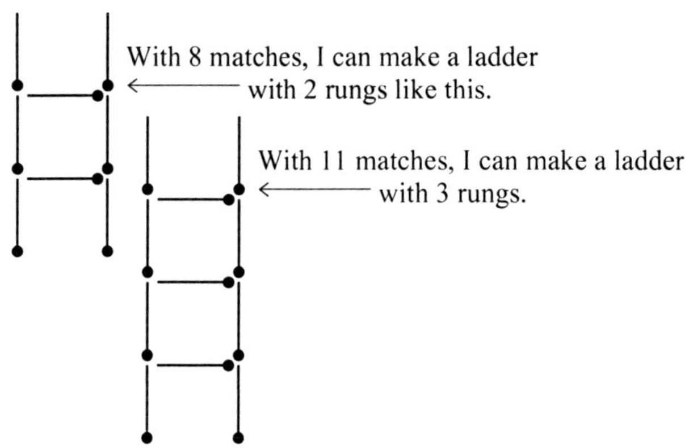

How many matches are needed to make the same sort of ladder with 4 rungs?
How many matches are needed to make a ladder with 5 rungs?
...

Figure 1-1. Ladders problem used by Stacey (1989, p. 148)

The most frequent error type on this problem was due to an assumption of proportionality instead of a determination of the correct affine relation between the number of rungs and the number of matches ('number of matches' = 3 × 'number of rungs' + 2). Students made use of the $f(kx) = k\,f(x)$ property of linear functions (e.g. "for a ladder with 4 rungs, one needs twice as many matches as for a ladder with 2 rungs"), of the $f(x + y) = f(x) + f(y)$ property (e.g., "one needs 8 + 11 matches for a ladder with 5 (= 2 + 3) rungs"), or they used combinations of both properties. Analogous observations of students' 'proportional multiplication error' (to use the term introduced by Linchevski, Olivier, Sasman, & Liebenberg, 1998) for similar pattern generalization problems are found in numerous other studies (e.g., Kuchemann & Hoyles, 2001; Lin & Yang, 2004; Linchevski et al., 1998; Orton & Orton, 1994).

Another case relates to the solving in university students involved in a first calculus course of problems like "If a plant measures 30 cm at the beginning of an experiment, and its height increases 50% monthly, how much will it measure after 3 months?" It was observed that 62% of the students reasoned linearly regarding the increase of the height as a function of time, instead of taking into account the exponential character of this growth process (Esteley, Villarreal, & Alagia, 2004; Villarreal, Esteley, & Alagia, 2004). Linear solutions were given even when a non-linear model (quadratic, exponential, logarithmic, etc.) was explicitly given in the problem statement by means of an analytic expression. For instance, for a problem like the one described in Figure 1-2, 10% of the students first calculated the height of the tree after six months with the given formula, but then used this result to incorrectly proceed with the 'rule of three' to answer the second question (Esteley et al., 2004; Villarreal et al., 2004). Probably, in applying the 'rule of three', students were not aware that they were applying the $f(kx) = k\,f(x)$ property, which only holds for linear functions.

Similar 'linearity errors' are reported in the manipulation of algebraic expressions. Every high school teacher can recall examples of students applying 'properties' like: "The square root of a sum is the sum of the square roots", or "the logarithm of a multiple is the multiple of the logarithm". This type of error has been discussed and illustrated by Gagatsis and Kyriakides (2000) and Matz (1982), but we are not aware of any empirical research. According to Matz (1982), these linearity errors result from students' overgeneralisation of the distributive law for the addition and multiplication of numbers to operations with non-linear functions (or with the symbols representing these functions). The immense number of occasions wherein students add and use the distributive law in arithmetic and early algebra is very likely to reinforce students' acceptance of linearity.

> After the first month of life, the growth of a tree follows the equation
>
> $$h(t) = 8 \cdot \log_2 t + 70$$
>
> where the height is given in cm and the time is given in months.
>
> a) Calculate the height of the tree after 6 months.
>
> b) Find the time it will take to the tree to reach a height of 1 m.

Figure 1-2. Problem described by Villarreal et al. (2004)

3.5 The overuse of linearity in numerical estimation

Because linear relations are essential to interpret many real world phenomena, it is not surprising that linearity is strongly present in our mathematical system itself. Even our number system and the activity of counting are expressions of linearity. The natural numbers are located at constant intervals starting from 0, which implies that 2 is twice as large as 1, that 100 consists of 5 equal parts of 20, etc. And when counting objects, the number counted is proportional to the number of 'count-events' and to the duration of counting. Initially, children do not necessarily experience the number system as linear. They may experience an equal space between 0 and 10 and between 10 and 20, but they may simultaneously experience a smaller distance between 12 000 and 12 500 than between 10 and 100 (see, e.g., Dehaene, 1997). This is shown in experiments where children were asked to locate these numbers on a number line where only the ends (e.g. 0 and 10 000) were labelled. Siegler and Opfer (2003) have demonstrated that throughout primary education, children gradually become more accurate in locating numbers on such an 'empty' number line segment and acquire the appropriate linear representation of the size of numbers.

But, interestingly, while this linear representation of the number system is being acquired, its overgeneralisation starts to occur as well. In a study on the development of the skill of estimating the magnitude of exponential expressions (Sastre & Mullet, 1998), students between 12 and 18 years old were confronted with expressions of the form a^n. Again, students were asked to locate the answer on an empty line segment. About half of the students in the total sample (and mainly the 12- to 16-year olds) exhibited a linear instead of an quadratic or exponential representation model in their estimations, for example, by choosing an equal distance between 2^2, 4^2, and 6^2, and also between 5^2, 5^4, and 5^6. Students here again may have applied the $f(kx) = k\,f(x)$ property in a non-linear context. But this phenomenon may also be explained by students' familiarity with the representation of number lines being divided in equal intervals (which is due to the linear character inherent to the number system itself).

4. THE CASE OF GEOMETRY

Probably the best-known cases of students' over-reliance on the linear model are situated in the domain of geometry. For instance, it is reported that incorrect linear reasonings frequently occur in problems about the relations between angles and sides of geometrical figures (see, e.g., Bold, 1969; De Block-Docq, 1992; Rouche, 1992a). According to Rouche (1992a), many students and even adults believe that an angle can be bisected or trisected by applying the constructions suggested in Figure 1-3, both based on improper linear assumptions.

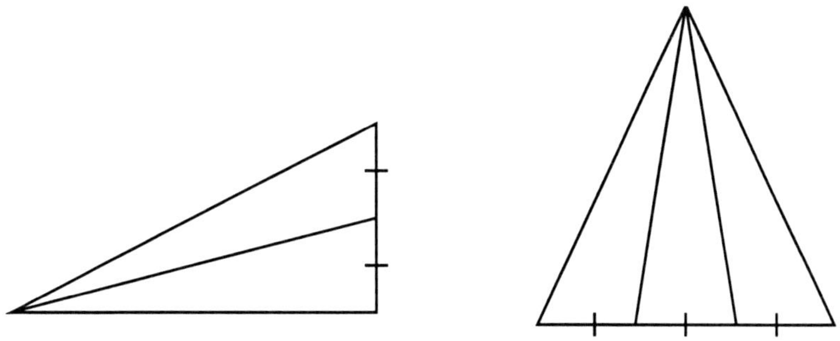

Figure 1-3. Constructions for the bisection and the trisection of an angle based on improper linear assumptions

The first construction assumes a linear relation between an acute angle and its opposite side in a right-angled triangle. The second construction, which is correct for the bisection but not for the trisection of an angle, assumes a linear relation between the angle at the top and the base of an isosceles triangle (or between an angle in a circle and its corresponding chord). Of course, in these specific cases, students can easily be disabused by confronting them with an application of these constructions on larger angles (Rouche, 1992a). Other examples of the overuse of linearity in the same domain of geometry can be found in the doctoral dissertation of De Block-Docq (1992). In the course of her "Epistemological comparative analysis of two teaching methods for plane geometry with pupils of the age of twelve", the author discusses several typical erroneous reasoning processes all based on the inappropriate application of direct (or inverse) proportionality between non-proportionally related magnitudes (e.g., "The angle of a regular dodecagon can be obtained by dividing the angle of a regular hexagon by six and multiplying this result by twelve", "If one can split a heptagon into five triangles (by joining one vertex to all the other vertices), one can split up a 14-sided polygon into ten triangles", "To construct an equilateral triangle inscribed in a circle, one has to step out the diameter on the circumference; an inscribed regular dodecagon can be constructed by stepping out half of the radius on the circumference", De Block-Docq, 1992, p. 199). In the last example, students' reasoning incorrectly infers from the fact that one can construct a regular hexagon in a circle by stepping out the radius. De Block-Docq (1992) only noticed linearity-based errors in which at least one of the variables in the assumed proportional relation is a discrete one (as in the above-mentioned examples). When two continuous variables were involved, the students rather tended to improperly assume an additive relation. For instance, they assumed the constancy of a difference, as in the following example: "To determine the height of an equilateral triangle, one has to remove 0.5 cm of the length of the side".

4.1 The effect of a linear enlargement (reduction) on area and volume

The rest of this monograph comprises a report and discussion of a series of closely related studies on students' solutions of application problems about the relation between the linear measurements of similar geometrical figures, on the one hand, and the area or the volume, on the other hand. The principle governing that kind of application problems is well known: A linear enlargement or reduction by factor r, multiplies lengths by factor r, areas by factor r^2 and volumes by factor r^3. A crucial aspect of understanding this principle is the insight that these factors depend only on the dimensionality of the magnitudes involved (length, area, and/or volume), and not on the particularities of the figures (whether these figures are squares, circles, etc.). As argued by many authors (e.g., Modestou, Gagatsis, & Pitta-Pantazi, 2004; Mogensen, 2004; National Council of Teachers of Mathematics, 1989; Outhred & Mitchelmore, 2000; Rogalski, 1982; Rouche, 1989; Simon & Blume, 1994; Tierney, Boyd, & Davis, 1990), gaining insight into the above-mentioned mathematical relations between lengths, areas, and volumes of similar figures, is usually a slow and painful process, whereby students are often misled because they assume linear relations. In this respect, we recall the famous example of the slave in Plato's dialogue *Meno*, who was asked to draw a square having two times the area of a given square and first proposed to double the side of that square (see, e.g., Berté, 1992, 1993; Daumas, 1989; Lelouard, Mira, & Nicolle, 1989; Rouche, 1989, 1992b). Or recall the legend of the Athenians – mentioned at the beginning of this chapter – doubling all sides of their altar at the request of the oracle of Apollo (Smith, 1923). More recently, we read in the *NCTM Standards* that "... most students in Grades 5–8 incorrectly believe that if the sides of a figure are doubled to produce a similar figure, the area and volume also will be doubled" (National Council of Teachers of Mathematics, 1989, pp. 114–115). In other words, students strongly tend to see the relations between length and area or between length and volume as linear instead of, respectively, quadratic and cubic, and, consequently, they apply the linear scale factor instead of its square or cube to determine the area or volume of an enlarged or reduced figure.

According to Freudenthal (1983), the above-mentioned principle governing the enlargement (or reduction) of geometrical figures is most fundamental in mathematics and science and, therefore, deserves our closest attention, both from a phenomenological and a didactical point of view. Remarkably, students' former real-life practices with enlarging and reducing operations do not necessarily make them aware of the different growth patterns of lengths, areas and volumes. In this respect, Feys (1995) describes the following experience with his student-teachers who were asked what will happen when they lay out two A4 pages side by side on a copier in order to reduce them on one A4 page. Regularly, they answered that the text will no longer be readable because the height and width of the characters and of the drawings will be halved. Another illustration is that students typically are surprised that in case of an enlargement, the area and certainly the volume is enlarged *that much*; and, when performing a reduction, they are often surprised that the area and certainly the volume is reduced *that much*. Indeed, a giant being ten times as tall as an adult man of 70 kg, would weigh 70 ton; a goblin ten times smaller than this adult, would only weigh 70 g (Streefland, 1984; Treffers, 1987)! A well-known example of students' misjudging of the effect of a reduction on volume is the half-filled conic glass: Students most often realise that its volume is less than the half of a full glass, but when asked to estimate more precisely which part it is, their answers are mostly more than one eighth. This overestimation is probably related to visual perception: In front view, one can see in fact a triangle which area is reduced to one fourth (see Figure 1-4).

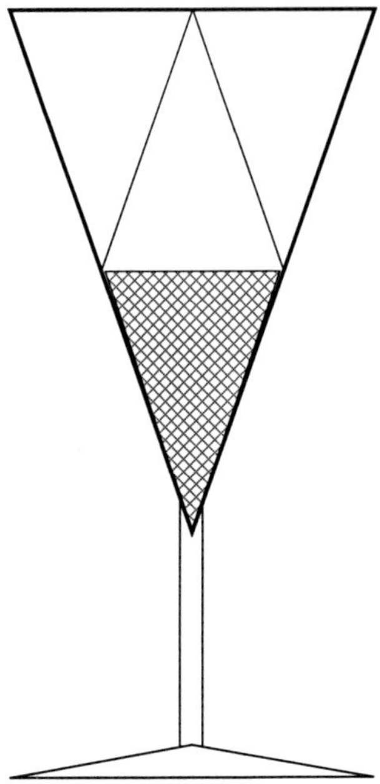

Figure 1-4. Front view of a half-filled conic glass

Finally, it is reality itself that, in a certain sense, sometimes puts the students on the wrong track. The mathematical idea of a 'linear enlargement' does not always fit the physical and biological reality of scaling (see, e.g., Haldane, 1928; Kindt & de Lange, 1986; Thompson, 1961). Old trees are plumper than younger specimens; tigers have relatively thicker paws than cats; the wings of an eagle are comparatively larger than those of a swallow; small mammals must keep on eating to stay warm. Also babies are not 'linearly reduced' adults. Their head and bones, in relation to their bodies as a whole, are relatively heavier than adults. These examples of enlargements 'taken from nature' are not similar enlargements. Lengths are not enlarged or reduced in all dimensions by the same factor. The reasons for this non-similarity are not mathematical, but have a physical or biological origin. Let us explain two of the above-mentioned examples. First, why must higher trees have relatively thicker trunks than smaller species? Suppose that the trunk of a tree twice as high was twice as thick. Then, the higher tree's volume would be increased by a factor of 8! In fact, the bearing-power of a trunk (pillar, paw, ...) is directly proportional to the cross-section of the trunk, so that, in the case of a linear enlargement by a factor 2 (in all dimensions), it would only quadruple. In order to support a tree 8 times as heavy, the diameter of the trunk should increase by a factor of $\sqrt{8}$ (which is nearly a triplication!). Second, why must small mammals keep on eating to stay warm? When the size of a body shrinks, its surface area reduces by the square of the scaling factor, while its volume reduces by the cube of that factor. Therefore, smaller bodies have, relative to their volume, larger surface areas than larger bodies of the same shape. For mammals and birds, this implies that smaller species lose their heat (through their skin) relatively faster, and thus must keep on eating to stay warm. We have no difficulty understanding and extrapolating linear increase or decrease of length, but we are easily defeated when the impact on ratios of surface area to volume are concerned.

4.2 Introduction to the research program

In chapter 2 we report two exploratory studies. A first study mainly investigated the occurrence and the strength of students' tendency to overuse linearity at the age of 12–13 when working on word problems about the effect of a linear enlargement or reduction of a geometrical figure on the perimeter or area of that figure. Moreover, the impact of two task variables was investigated: (1) the shape of geometrical figure involved in the problem situation and (2) the availability of self-made or ready-made drawings accompanying the problem statement. The second study was basically a replication of the first one with 15–16-year old students.

Chapter 3 contains a description of a series of follow-up studies wherein we investigated the effects of different modifications of the experimental context, to test different hypothetical explanations for this phenomenon. In a first follow-up study, we altered the problem presentation by including metacognitive and visual scaffolds aimed at arousing students' doubts about the appropriateness of the linear model and at helping them to find the appropriate mathematical model. In a second follow-up study, we transformed the problems, which were originally formulated in a missing-value format, into comparison problems. Third, we investigated the impact of (1) linking word problems to an authentic context – acting on students' intrinsic motivation – and (2) self-made graphical presentations integrated in the problem solving with a view to help students to make a correct situational model of the problem. Finally, we reformulated items so that they expressed perimeter and area by direct measures instead of indirect measures (as was the case in previous studies). Generally speaking, all these experimental manipulations had positive effects on students' ability to discover the (in)appropriateness of the linear model, but these effects remained rather small, indicating that students' tendency towards linear modelling is very strong, deep-rooted and resistant to change.

In chapter 4 we make a methodological shift from collectively testing large groups of students to individual interviews with smaller numbers of students. More concretely, we report an in-depth investigation aimed at unravelling the actual process of problem solving from students falling into the 'linearity trap' and the mechanisms behind it. The results of this study enabled us to identify the role of different aspects of students' knowledge base that were responsible for their inappropriate proportional responses.

Chapter 5 contains the report of a study in which an experimental lesson series was developed and tested in the context of a design experiment. The main goal was to investigate whether it is possible to develop in lower secondary students the necessary knowledge elements, skills, and attitudes to respond adequately to linear and non-linear application problems about the perimeter, area and volume of enlarged geometrical figures (and thus to break the tendency to overuse the linear model in this curricular domain).

In chapter 6 we step outside the classroom context. We report a study conducted to see whether a more drastic change of the experimental or instructional setting (namely offering a non-linear problem as a meaningful, authentic performance task instead of a traditional, school-like word problem as in all the previous studies) could break the overuse of linearity. This study also allowed us to investigate whether the experience of solving such an authentic performance task influences students' later behaviour when they solve school-like word problems that deal with the same issue.

Finally, in chapter 7 we discuss the psychological and educational factors that lie at the roots of the occurrence and persistence of the over-reliance on linearity. Explanatory factors were found in (1) the intuitive, heuristic nature of the linear model, (2) students' experiences in the mathematics classroom, and (3) elements related to the mathematical particularities of the problem situation in which the linear error occurs. We conclude that final chapter with some recommendations for the improvement of educational practice. First, we suggest some modifications with respect to the mathematics curricula and textbooks. These modifications would already constitute a first instructional response to students' tendency towards improper linear reasoning. Afterwards, we formulate a more comprehensive response, which consists in bringing the modelling perspective more to the forefront of mathematics education and, consequently, shifting the focus from computing correct numerical solutions to building appropriate mathematical models.

Chapter 2

IN SEARCH OF EMPIRICAL EVIDENCE

1. INTRODUCTION

Notwithstanding the numerous illustrations of students' tendency towards the overuse of linearity, systematic observations and analyses of this phenomenon based on empirical research have been absent until recent years. In this chapter, we report a first pair of studies aimed at filling this gap in the research literature (see also De Bock, Verschaffel, & Janssens, 1998). Both studies focused on application problems about the effect of a linear enlargement or reduction of a geometrical figure on the perimeter or area of that figure. As illustrated in the previous chapter, many scholars in the field have mentioned the occurrence of students' overuse of linearity for this particular type of problem situation, most often incidentally in the course of studies with other main foci (e.g. Mogensen, 2004; Outhred & Mitchelmore, 2000; Rogalski, 1982; Simon & Blume, 1994; Tierney, Boyd, & Davis, 1990).

A first study investigated the occurrence and the strength of 12–13-year old students' tendency to overuse linearity in solving word problems about the effect of a linear enlargement or reduction of a geometrical figure on the perimeter or area of that figure. Moreover, the impact of two task variables was investigated: (1) the shape of the geometrical figure involved in the problem situation, and (2) the availability of self-made or ready-made drawings accompanying the problem statement. Due to the extremely low occurrence of correct answers in this study, however, it was hardly possible to determine the impact of these task variables. Therefore, a second study was conducted, which was basically a replication of the first one, but now with an older age group of 15–16-year old students.

2. STUDY 1

2.1 Subjects, materials and procedure

Participants in the first study were 120 12–13-year olds (7th-graders) of a large secondary school in Flanders recruiting its students from a wide range of elementary schools in the region. Therefore, although all students attended the same secondary school at the time of testing, their educational histories with respect to (elementary) mathematics education were not identical but there are common elements. Word problems dealing with direct proportionality constitute an important topic of the mathematics program in the upper grades of the primary school in Flanders (Ministerie van de Vlaamse Gemeenschap, 1997). With the exception of inverse proportionality ($y = k/x$), other types of functional relationships are normally not systematically addressed at this level of schooling. With respect to elementary geometry, Flemish students are expected to master the names and the basic properties of the most familiar geometric figures, including the formulas for calculating their perimeter, area and volume, by the end of elementary school. Their expected experience with scaling and similarity in relation to mensuration involves declarative and procedural knowledge of the rules of the metric system for one-, two-, and three-dimensional measures.

In this first study, the 120 students were divided into three equivalent groups (Groups I, II, and III consisting of 40, 42, and 38 students, respectively). Each group was composed of two intact classes, one of which had six hours of mathematics a week, while the other class had four hours of mathematics a week.

The experiment consisted of two phases. During the first phase all 120 students were administered the same paper-and-pencil test consisting of 12 experimental items and 3 buffer items (Test 1). No clues or special instructions were given. All 12 experimental items involved enlargements of similar plane figures, and belonged to three categories: 4 items about *s*quares (S), 4 about *c*ircles (C), and 4 about irregular figures (I). Within each category of figures (S, C, and I), there were 2 proportional and 2 non-proportional items. Table 2-1 lists examples of one proportional and one non-proportional item for each of the three categories of figures from Test 1.

Table 2-1. Examples of experimental items

Enlargement of a square figure

Proportional item:

Farmer Gus needs approximately 4 days to dig a ditch around a square pasture with a side of 100 m. How many days would he need to dig a ditch around a square pasture with a side of 300 m? (Answer: 12 days)

Non-proportional item:

Farmer Carl needs approximately 8 hours to fertilise a square piece of land with a side of 200 m. How many hours would he need to fertilise a square piece of land with a side of 600 m? (Answer: 72 hours)

Enlargement of a circular figure

Proportional item:

You need approximately 6 hours to sail around a circular island with a diameter of 70 km. How many hours would you need to sail around a circular island with a diameter of 140 km? *(Answer: 12 hours)*

Non-proportional item:

You need approximately 400 grams of flower seed to lay out a circular flower bed with a diameter of 10 m. How many grams of flower seed would you need to lay out a circular flower bed with a diameter of 20 m? *(Answer: 1 600 grams)*

Enlargement of an irregular figure

Proportional item:

On a map of Belgium in an atlas the distance from Genk to Leuven is approximately 5 cm and the distance from Genk to Ghent approximately 11 cm. On a map in front of the classroom the distance from Genk to Leuven is approximately 20 cm. How long is the distance from Genk to Ghent on this map? *(Answer: 44 cm)*

Non-proportional item:

On a map of Belgium in an atlas the distance from Genk to Tongeren is approximately 2 cm and the area of Belgium approximately 250 cm^2. On a map in front of the classroom, the distance from Genk to Tongeren is approximately 6 cm. How large is the area of Belgium on this map? (Answer: 2 250 cm^2)

As illustrated in Table 2-1, the variables 'length' and 'area' were mostly replaced by more concrete, indirect variables that are proportional to them (or are reasonably supposed to be so), with a view to constructing a set of meaningful application problems. We come back to this issue in chapter 3. All other possibly relevant task variables – such as the degree of familiarity with the problem context, the grammatical complexity of the problem formulation, and the nature of given numbers – were controlled as much as possible. For instance, we used only 'simple' natural numbers as scale factors, so that all required computations had a similar technical difficulty.

The response sheets could be used not only to write the answers, but also to make calculations, drawings, or comments.

Two weeks after the first test the three groups of students were confronted with a second test (Test 2), which was a parallel version of Test 1. Once again, the problems were the same in all three groups, but the way in which the test was introduced and presented was different. In Group I, which functioned as the control group, the testing conditions were exactly the same as during the first test, i.e., these students received the problems without any additional clues and in exactly the same format as during Test 1. The students of Group II were explicitly instructed to make a sketch or drawing before answering each problem. This instruction was given at the beginning of the test and was illustrated by means of an example item (which did, of course, not involve similar plane figures). In Group III every problem was accompanied by a relevant ready-made drawing like the one given in Figure 2-1 (belonging to the non-proportional item about squares in Table 2-1).

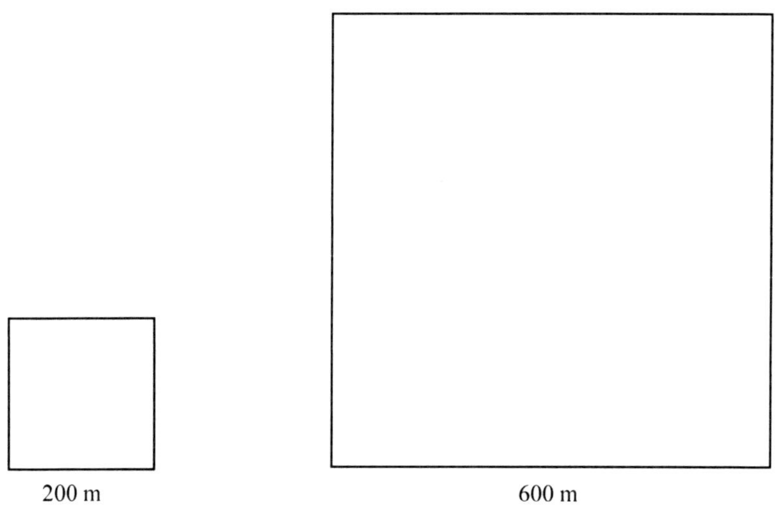

Figure 2-1. Example of a ready-made drawing

2.2 Hypotheses

First, on the basis of what is generally acknowledged in the field, we hypothesized that the predominance of the linear model would be a serious obstacle for the vast majority of the students. Consequently, we predicted that their performance on the 6 proportional items would be very high, while their scores on the 6 non-proportional items would be very low.

Second, we hypothesized that a sketch or drawing would have a beneficial effect on the students' performance, especially for the non-proportional experimental items. This hypothesis is based on the vast amount of theoretical and empirical research on how and why drawings and diagrams are a useful in enhancing people's ability to represent and solve (mathematical) problems (Aprea & Ebner, 1999; De Corte, Greer, & Verschaffel, 1996; Hall, Bailey, & Tillman, 1997; Larkin & Simon, 1987; Pólya, 1945; Reed, 1999; Schoenfeld, 1992; Vlahovic-Stetic, 1999). When students are asked to *make a drawing themselves*, they are stimulated to construct a proper (mental) representation of the essential elements and relations involved in the problem (Pólya, 1945; Schoenfeld, 1992). Especially for the non-proportional items, this representational activity should help students to detect the inappropriateness of a stereotyped linear reasoning, and to determine the nature of the non-linear relationship connecting the known and the unknown elements in this problem representation. Of course, the heuristic of making a drawing or diagram does not guarantee that one will find the solution of a given problem. For instance, a drawing that reflects an incorrect understanding of the problem will be of little help for the problem solution (Van Essen & Hamaker, 1990). When students do not succeed in making a *correct, usable* drawing themselves, it might be more effective if they are *provided with a correct ready-made drawing*. Starting from these hypotheses, we predicted that in Group I the results will be the same for Test 1 and Test 2, while in Group II and Group III, the percentage of correct responses will increase from Test 1 to Test 2. This increase would be essentially due to a decrease of inappropriate solutions based on linear reasoning on the non-proportional items during Test 2. No specific hypothesis was stated with respect to the relative impact of the two experimental manipulations.

Third, we predicted that students' performances would be different for the distinct types of plane figures involved in the study. More specifically, the items about squares (S items) were supposed to be the easiest and those about irregular figures (I items) the most difficult. We also expected that the size of the anticipated effect of drawings (see hypothesis 2) would be affected by the type of figure, in the sense that this drawing effect would be the greatest for the (non-proportional) S items and the lowest for the (non-proportional) I items. The rationale behind these predictions is exemplarily worked out for the non-proportional item about squares given in Table 2-1. To find the answer to this item, the problem solver can choose among three appropriate solution strategies: (1) 'paving' the big square with little ones, (2) calculating and comparing the areas of both squares by means of the formula 'area = side × side', and (3) applying the general principle 'if length × r, then area × r^2'. For the corresponding non-proportional item about circles, the first solution strategy is less obvious and can only provide an approximate answer, while the second strategy is more error-prone (because of the greater complexity of the formula for finding the area of a circle). For the corresponding non-proportional item dealing with irregular figures, a formula for calculating the area is not available and the paving strategy – or a variant, namely approximately transforming the irregular figure into one or more regular ones – does not provide an 'exact' answer. Applying the general principle is the only 'direct' solution strategy for this item.

2.3 Analysis

All responses on the proportional and the non-proportional items were categorized as 'correct' or 'incorrect'. A response was considered correct when it was the result of a mathematically appropriate reasoning process; therefore, answers that differed from the correct one because of a technical mistake in a correct overall solution process were considered correct too. All other kinds of erroneous answers were scored as incorrect. Because a detailed analysis of a random sample of 300 incorrect answers on non-proportional items revealed that 95% of them resulted from an inappropriate linear reasoning process (see De Bock, Verschaffel, & Janssens, 1996), we decided not to split up the incorrect answers any further in the present analysis.

The hypotheses were tested by means of a '3 × 2 × 2 × 3' analysis of variance (ANOVA) with 'Group' (Group I, II, and III), 'Test' (Test 1 vs. Test 2), 'Proportionality' (proportional vs. non-proportional items), and 'Figure' (squares, circles, and irregular figures) as independent variables, and the number of 'Correct answers' as the dependent variable. Significant main and interaction effects were further analyzed using a posteriori Tukey tests.

2.4 Results with respect to the hypotheses

Table 2-2 gives an overview of the percentages of correct responses of the three groups of students (I, II, and III) on the proportional and the non-proportional problems involving squares (S), circles (C), and irregular figures (I) in Test 1 and 2.

The results provide a very strong confirmation of the first hypothesis. Indeed, there was a strong difference in the performance on the proportional and non-proportional problems[4]. For the three groups and the two tests together, the percentages of correct responses for all proportional and for all non-proportional items were 92% and 2%, respectively. This shows that most 12–13-year olds were able to solve the proportional items correctly, whereas the non-proportional items were seldom solved correctly.

The results did not support the second hypothesis concerning the beneficial effect of drawings on students' performance. For none of the three groups did we find a significant increase in the students' scores from Test 1 to Test 2 in general, or in their performance on the non-proportional items in particular. For Group I (= the control group) the percentage of correct responses on the non-proportional items even decreased slightly from 2% to 1% between Test 1 and Test 2. For Group II (= the self-made drawing group) this percentage unexpectedly remained the same at 2% during both tests. For Group III (= the ready-made drawing group) the percentage of correct answers on the non-proportional items increased slightly between Test 1 and Test 2, from 2% to 5%, but remained still extremely low during the latter test. In sum, the anticipated beneficial impact of the instruction to make drawings (in Group II) and of the provision of ready-made drawings (in Group III) was too weak to break the predominance of the linear model in the reasoning of these 12–13-year olds.

In line with the third hypothesis, the type of figure had a significant effect on the percentage of correct responses[5]. The scores for the S, the C, and the I items were in the expected direction – the overall percentages of correct answers for these three kinds of problems were 49%, 48%, and 45%, respectively, but only the difference between the S items and the I items and between the C items and the I items was statistically significant. Moreover, the observed effect of the type of figure was found in the proportional items as well as in the non-proportional items.

Table 2-2. Percentage of correct responses of the three groups of 12–13-year olds on the different categories of proportional and non-proportional items in Test 1 and Test 2

Group	Test 1						Test 2					
	Proportional items			Non-proportional items			Proportional items			Non-proportional items		
	S	C	I	S	C	I	S	C	I	S	C	I
I	96	98	89	5	0	1	99	96	85	3	0	0
II	93	95	89	6	1	0	93	95	95	4	2	0
III	91	91	87	4	3	0	93	89	89	8	5	1

4 'Proportionality' main effect: $F(1,117) = 4994.92$, $p < .01$

5 'Type of figure' main effect: $F(2,234) = 12.96$, $p < .01$

2.5 Additional findings

After presenting the quantitative results with respect to the three research hypotheses, we also briefly discuss some qualitative findings based on a systematic and fine-grained analysis of the students' written notes on the response sheets, which may help to explain these quantitative results.

First, the analysis of the notes on the response sheets of Test 1 revealed that only 2% of the students *spontaneously* constructed a sketch or drawing of the non-proportional problems. Apparently, these 12–13-year olds were not inclined to apply the heuristic of making a sketch or a drawing when modelling and solving a verbally stated geometric problem situation.

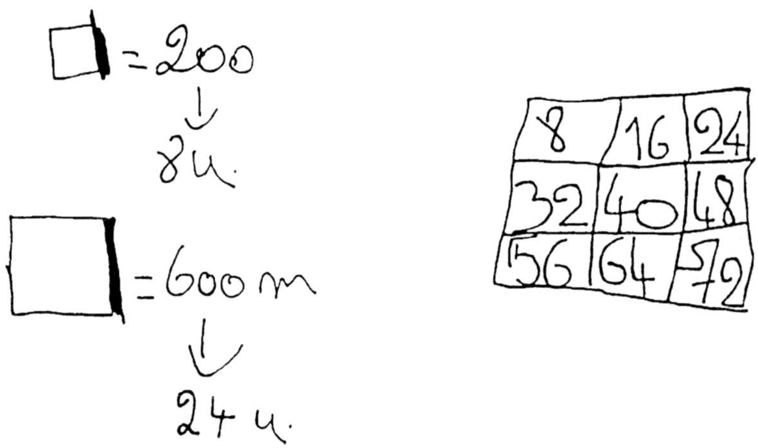

Figure 2-2. Examples of self-made drawings with low (left) and high (right) representational quality

Second, the inspection of the notes on Test 2 in Group II revealed that – in spite of the explicit instruction to do so – the students produced drawings for the non-proportional items in only 46% of the cases. It remains unclear why even the students from Group II who *did* follow the instruction to make a drawing still failed to solve the corresponding non-proportional item. Possibly, the representational quality of these drawings (both in terms of correctness and richness) was mostly too low to really help students in interpreting and solving these non-proportional items correctly. For examples of self-made drawings with a low and a high representational quality, we refer, respectively, to the drawings left and right in Table 2-2, both belonging to the non-proportional item about squares in Table 2-1. On the basis of the available data we cannot determine whether the poor quality of most student drawings was due to their inability to make better drawings or to their unwillingness to make drawings for word problems that seemed trivial to them.

Third, although the notes of the students from Group III on Test 2 provide no *direct* information about the extent to which the ready-made drawings were effectively used by these students, the finding that only 6% of the given drawings for the non-proportional items had been ‘edited’ by these students as part of their problem-solving process, suggests that they generally paid little or no attention to them.

2.6 Conclusion and discussion

The first study convincingly demonstrated the strength and omnipresence of the linear model with respect to problems involving length and area of similar plane figures in 12–13-year old students. However, for two reasons we found it necessary to set up a follow-up study with an older target group.

First, the extremely small number of correct responses on the non-proportional items made us wonder how strong the predominance of linearity would be for students who were older and, therefore, probably mathematically better equipped for overcoming the obstacle of unlimited linear reasoning.

Second, because the predominance of the linear model proved to be so strong for 12–13-year olds, the first study did not yield adequate information about the possible impact of both the kind of figure used in the problem (square, circular or irregular) and of self-made or ready-made drawings on the occurrence of errors based on inappropriate proportional reasoning. For instance, because of the extremely low occurrence of correct answers on the non-proportional items in Study 1, it was impossible to perform a statistical analysis of the relationship between making and/or using drawings, on the one hand, and producing correct answers, on the other hand. A follow-up study with older students – who were expected to suffer less from the predominance of the linear model – should result in a better understanding of the effect of these two additional task variables.

3. STUDY 2

3.1 Subjects, materials and procedure

For this follow-up study, we decided to work with 10th-graders (15–16-year olds). This choice was induced by the Flemish mathematics program for

Grades 9–10 (Ministerie van de Vlaamse Gemeenschap, 2002) which includes a systematic study of plane geometry (including similarities and scaling in relation to mensuration). Furthermore, in Grades 9 and 10 students also learn how to identify and use a diversity of non-proportional functional relationships (e.g. quadratic and general polynomial functions).

To work with a cohort of participants of an intellectually, educationally and socially comparable level to that of the first study, we decided to undertake the second study in the same school as the first one. Participants were the 222 10th-graders (15–16-year olds).

Contrary to the first study, the test was administered only once to all students. Therefore, students were immediately matched in three equivalent groups. For practical reasons, it was not possible to match the students from the three experimental groups on an individual basis; rather, we had to work with intact classes. The selection and placement of the distinct classes in the three experimental groups was based on the following subject variables: (1) the study stream to which the students of a certain class belonged (e.g., 'Ancient languages', 'Modern languages', 'Economics', ...), (2) the number of hours a week spent at mathematics, and (3) the students' results on the previous mathematics examination. Using these available data, three groups were formed in which (1) the different study streams were equally represented, (2) the average number of hours of mathematics per week was the same, and (3) the average result on the previous mathematics examination was similar (for a detailed description of the characteristics of the three experimental groups, see De Bock et al., 1996). The testing conditions for the three experimental groups were the same as in Test 2 of Study 1, i.e. in Group I no special help or instructions were given, in Group II students were explicitly instructed to make a drawing before computing their answer, and in Group III every item was accompanied by a correct ready-made drawing.

The same 12 experimental items from the first study (Table 2-1) were used. The testing procedure, the layout of the response sheets, and the data-analysis procedure were also identical.

3.2 Hypotheses

The predictions of Study 2 were largely the same as for Study 1. However, with regard to the first hypothesis, we assumed that 15–16-year olds would suffer less from the predominance of the linear model than 12–13-year olds. Therefore, we predicted that they would perform better on the experimental items in general and on the non-proportional ones in particular.

The hypotheses regarding the influence of self-generated or ready-made drawings (= hypothesis 2 from Study 1) and of the type of geometrical figure involved (= hypothesis 3 from Study 1) remained unchanged.

3.3 Results with respect to the hypotheses

Table 2-3 lists the percentages of correct responses for the three groups of 15–16-year olds (I, II, and III) for the proportional and non-proportional items involving squares (S), circles (C), and irregular figures (I).

Table 2-3. Percentage of correct responses of the three groups of 15–16-year olds on the different categories of proportional and non-proportional items

Group	Proportional items			Non-proportional items		
	S	C	I	S	C	I
I	91	97	97	26	11	1
II	89	91	97	26	20	5
III	89	93	97	39	21	7

The first hypothesis was confirmed. Once again, there was an extremely strong difference in performance on the proportional and non-proportional items[6]. For all three groups together, the percentages of correct responses on the proportional and non-proportional items were 93% and 17%, respectively. (In Study 1, these percentages were 92% and 2%). A comparison of the results from the two studies revealed that only on the non-proportional items was the difference in the number of correct responses between the 7th- and the 10th-graders significant[7].

The second hypothesis about the positive influence of drawings was again not confirmed. Global percentages of correct responses in Groups I, II, and III showed indeed a positive trend – i.e. 54%, 55%, and 58% correct answers, respectively – but the differences were again too small to produce a significant effect.

As predicted in the third hypothesis and in accordance with the results of Study 1, the type of figure did make a difference[8]. Global percentages of correct responses for S, C, and I items were in the expected direction (60%, 56%, and 51% correct answers, respectively), and all mutual differences between these three problem types were significant. Interestingly, the percentages of correct responses on the non-proportional items were in the expected direction (30%, 17%, and 4% for the S, C, and I items,

[6] 'Proportionality' main effect: $F(1,219) = 1591.64, p < .01$

[7] 'Age' × 'Proportionality' interaction effect: $F(1,340) = 25.33, p < .01$

[8] 'Type of figure' main effect: $F(2,438) = 24.67, p < .01$

respectively), whereas the percentages of correct responses on the proportional items were in the opposite direction (90%, 94%, and 97% for the S, C, and I items, respectively).

3.4 Additional findings

Students' use of self-made and ready-made drawings. First, we will describe some additional findings concerning the use of drawings in students' solutions of the non-proportional items. According to hypothesis 2, the instruction to make a drawing and – even more – the provision of a ready-made drawing should help students to overcome the obstacle of unlimited linear reasoning. Therefore, we predicted better results (on the non-proportional items) for Groups II and III than for Group I. But, as said before, the results of Study 2 again yielded no confirmation of this prediction. The qualitative analysis of the response sheets of the students of Group II and III in Study 1 suggested that the instruction to make drawings – and even the given drawings – were often ignored by these students. Therefore, the absence of a significant difference between the three groups in both studies did not allow the conclusion that the *actual* making of a drawing and the *actual* use of a ready-made drawing did not have a positive influence on students' performance. To make that conclusion, we first needed to demonstrate that there was no relationship between *actually* making a drawing or *actually* using a ready-made one, on the one hand, and giving a correct response to a non-proportional item, on the other hand. Because of (1) the extremely low overall score on the non-proportional items and (2) the relatively small number of self-generated and 'edited' drawings in the first study, it was impossible to analyze (statistically) the relationship between students' *actual* use of these drawings and their performance on the non-proportional items. Study 2, however, did allow such an analysis, because of the greater number of (1) self-generated and 'edited' drawings for the non-proportional items and of (2) correct responses on these items. For each of the three groups (Group I, II, and III) a contingency table with the variables 'Drawing' and 'Answer' was constructed (Table 2-4), and the (in)dependence of the two variables was investigated.

Chi-square tests revealed that this null hypothesis must be rejected in all groups[9]. The comparison of the observed and expected frequencies in the various cells for Group I, II, and III shows that the dependence of the two variables is in the expected direction. In all three tables the number of subjects in the cells on the main diagonal (i.e., 'drawing/correct answer' and

[9] $\chi^2(1, N = 444) = 16.18$, 25.26 and 14.82 for Group I, II, and III, respectively

'no drawing/incorrect answer') is greater than could be expected if these two variables were mutually independent. To investigate if the application of a drawing indeed *provoked* the apprehension of non-linearity, or if – conversely – this heuristic was used more often by students who already had detected the non-linear nature of the mathematical model underlying the problem, we compared the occurrence of spontaneous drawings for proportional and non-proportional items. This analysis revealed that the incidence of these drawings was not significantly higher for non-proportional than for proportional items, which suggests that students' insight in the non-linear character of an item generally *came after* the (successful) application of the drawing heuristic.

Table 2-4. Contingency tables for Group I, II, and II (Frequencies and percentages between brackets represent the expected cell frequencies and percentages under the null hypothesis of independence of both variables)

Group I	Spontaneous drawing	No spontaneous drawing	Row totals
Correct answer	17 (7) 4% (2%)	41 (51) 9% (11%)	58 13%
Incorrect answer	40 (50) 9% (11%)	346 (336) 78% (76%)	386 87%
Column totals	57 13%	387 87%	444 100%

Group II	Drawing made	No drawing made	Row totals
Correct answer	65 (45) 15% (10%)	12 (32) 3% (7%)	77 17%
Incorrect answer	196 (216) 44% (49%)	171 (151) 39% (34%)	367 83%
Column totals	261 59%	183 41%	444 100%

Group III	Given drawing edited	Given drawing not edited	Row totals
Correct answer	32 (19) 7% (4%)	66 (79) 15% (18%)	98 22%
Incorrect answer	53 (66) 12% (15%)	293 (280) 66% (63%)	346 78%
Column totals	85 19%	359 81%	444 100%

Students' solution strategies. Because the 15–16-year olds in Study 2 produced considerably more correct responses on the non-proportional items than the 12–13-year olds from Study 1, it also became possible to get further insight into the kind of strategies underlying these correct answers. For this purpose, all solution processes underlying a correct answer on a non-proportional item were scored in terms of one of the following categories: (1) the 'paving' strategy, (2) the strategy of computing and comparing the length or area of both figures, and (3) the strategy of applying the general rule. For instance, the non-proportional item about squares in Table 2-1 can be solved in three different ways:

1. 'Paving' the big square with little ones, observing that there are 9 little squares, and then concluding that the farmer will therefore need 9 times 8 hours = 72 hours;
2. Calculating the area of both squares ($200 \times 200 = 40\,000$ m^2 and $600 \times 600 = 360\,000$ m^2), determining the result of the division ($360\,000 : 40\,000 = 9$), and therefore concluding that the farmer will need 9 times 8 hours = 72 hours; and
3. Immediately applying the general rule 'side $\times$ r, thus area $\times$ r^2' ('if the length is multiplied by 3, then the area (and thus the fertilising time) needs to be multiplied by 9').

A detailed analysis of the written protocols of all correct solutions revealed that the second solution strategy was – by far – the most frequent one. For instance, for the non-proportional items about squares, this strategy (sometimes used in combination with one of the other methods) was applied for 90% of the correctly solved non-proportional items. 'Pure' applications of the first and the third strategy were very rare (7% and 3% of the correct solution strategies, respectively). The low frequency of the 'paving' strategy is quite remarkable. Indeed, 'paving' is a very easy, intuitive, context-bound method requiring only little sophisticated formal-mathematical knowledge. The rare use of this informal strategy was probably due to students' well-documented belief that solving a mathematical problem is primarily a matter of finding and executing the correct mathematical formula(s) previously taught in school (Schoenfeld, 1992; Verschaffel, et al., 2000).

Students' errors on proportional items. With respect to the type of figure, the percentages of correct responses on the different kinds of non-proportional items were in the expected direction (i.e. highest score for the S items and lowest score for the I items), whereas for the proportional items the percentages of correct responses were in the *opposite* direction (i.e. a higher percentage of correct answers for the I items than for the S items). A detailed analysis of students' incorrect responses on the proportional items showed that they were typically the result of an inappropriate *non-proportional* reasoning process. Apparently, students found it easier to discover the non-proportional nature of a given problem situation when a square figure was involved (and to some extent also when the figure had a circular form), but, as a result of this discovery, they sometimes began to question the correctness of a linear model for problem situations wherein that model *was* appropriate. In this respect, we point out that also in Study 1 some of the rare students who produced correct answers on non-proportional items, erroneously applied non-proportional reasoning for one or more proportional items too.

4. DISCUSSION

The major results of this first pair of studies were as follows. First, the tendency to apply linear reasoning in the solution of non-proportional problems proved to be extremely strong in the age group of 12–13-year olds, and was still very influential among 15–16-year olds. The 12–13-year olds solved 92% of the non-proportional items correctly, whereas they only answered 2% of the non-proportional items correctly; in the age group of 15–16-year olds, the overall percentages of correct responses on the proportional and non-proportional items were, respectively, 93% and 17%. These data, which can be interpreted as strong evidence for students' overuse of linearity, elicited a lot of amazement and unbelief among practitioners to whom we presented the results of these studies. Most of them were aware of the issue, but had not realised that it affected their students' solutions so strongly.

Second, the type of figure played a significant role. Students performed significantly better on the non-proportional items when the enlarged figure involved was regular (a square or a circle), but, as a drawback, they performed worse on the proportional items about these regular figures, because some students started to apply non-proportional reasoning on the proportional items too! This latter finding seems to indicate that the knowledge base underlying students' (rare) correct answers on non-proportional items was typically still quite fuzzy and unstable. Third, we unexpectedly did not find a beneficial effect of the self-made or given drawings, either for the test as a whole or for the non-proportional items in particular. An additional qualitative analysis of the data revealed that the vast majority of correct responses on the non-proportional items were found by applying an appropriate mathematical formula; informal strategies, such as paving a self-made or a given drawing, were chosen far less. It appears that we did not succeed in integrating the drawing activity or the given drawings in students' problem-solving process.

While this first set of investigations documented students' strong tendency to apply linear reasoning in the solution of non-proportional problems, it did not yield an explanation for this tendency. One could argue that students' unbridled use of the linear model was just an artefact of certain elements in our experimental setting. In that case, it would be possible to improve students' performance on the non-proportional items by simply modifying these elements. In the next chapter, we report a series of follow-up studies, each testing one or more particular hypotheses with respect to the influence of the testing setting on students' tendency to overuse linearity.

Chapter 3

SEARCHING FOR EXPLANATIONS: A SERIES OF FOLLOW-UP STUDIES

1. INTRODUCTION

The two studies reported in the previous chapter provided strong empirical evidence for the strength of 12–13- and 15–16-year old students' tendency to overuse linearity in geometrical problems about the relation between the linear measurements and the area of similar plane figures. Considering these very pronounced results, we decided to undertake a series of follow-up studies to test several hypothetical explanations for the phenomenon. While we assumed that the observed tendency to overuse linearity was at least partly due to certain deficiencies in students' knowledge base, we claimed that the observed findings were also partly caused by the way in which the problems were formulated and presented to the students and by particularities of the testing setting and, as such, might be less dramatic than suggested above. Relying on the psychological literature on (mathematical) problem solving, we explored the impact of several critical adaptations or modifications of the problem formulation and presentation and of the testing setting on the number of unwarranted linear responses.

In a first follow-up study (see also De Bock, Verschaffel, & Janssens, 2002b), we tested the hypothesis that students reasoned proportionally simply because they did not invest sufficient mental effort in solving the problems. This hypothesis was tested by including a metacognitive scaffold aimed at arousing students' doubts about the appropriateness of the linear model and at helping them to find the appropriate mathematical model. We also tested whether the absence of a substantial effect of self-made or given drawings in the previous studies might be caused by the absence of useful reference points in these drawings. Therefore, we also tested the effect of providing ready-made drawings on squared paper (instead of plain paper as in the first set of studies reported in chapter 2).

In the second follow-up study (see also De Bock et al., 2002b), we tested whether students' unbridled use of proportionality was affected by the formulation of the word problems in a missing-value format, which students might have learnt to associate with the proportionality scheme. For this purpose, the problems (formulated in a missing-value format in all previous studies) were reformulated as comparison problems.

A third follow-up study (see also De Bock, Verschaffel, Janssens, Van Dooren, & Claes, 2003) investigated whether the inauthentic character of the context of the word problems that were used in the test also had an impact on students' improper proportional reasoning. This hypothesis was tested by embedding the problems in a rich and attractive context. Additionally, we analysed whether self-made graphical presentations had a beneficial effect when they were truly integrated in students' problem-solving activity.

Finally, in a fourth follow-up study (see also Van Dooren, De Bock, De Bolle, Janssens, & Verschaffel, 2003a), we investigated whether the use of items using indirect perimeter and area measures (as was the case in all previous studies) might have strengthened students' tendency to give linear responses, by comparing their performances and solution strategies on items with direct and indirect measures.

2. FOLLOW-UP STUDY 1: EFFECTS OF METACOGNITIVE AND VISUAL SCAFFOLDS

2.1 Design and rationale

Two hundred and sixty 12–13-year olds and 125 15–16-year olds participated in the study. In both age groups, students were divided into four equivalent subgroups based on an individual matching procedure. This matching was based on the following student variables: (1) the study stream to which the student belonged (e.g., 'Ancient languages', 'Modern languages', 'Economics', …), (2) the number of hours a week spent at mathematics, and (3) the students' results on the last mathematics examination. Using these available data, four groups were formed in which the different study streams were equally represented and for which the average number of hours of mathematics per week and the average result on the last mathematics examination were similar. All four subgroups received the same paper-and-pencil test as in the studies reported in chapter 2 (see Table 2-1), but the administration of the test was different for each group. In the M–V– group (the no metacognitive and no visual scaffold group), no special help was given.

In the M+V– group (the metacognitive scaffold and no visual scaffold group), the test was preceded by an introductory task involving the following non-proportional item about a cube: "A wooden cube with an edge of 2 cm weighs 6 grams. How heavy is a wooden cube with an edge of 4 cm?" This item was accompanied by two alternative solution strategies presented as the answers of two fictitious peers.

> Two pupils, Wim and Steve, give a different solution to that problem.
>
> Wim argues: 4 cm is twice as long as 2 cm, thus I have to multiply the weight by 2. So, I do 6 grams × 2 = 12 grams.
>
> Steve argues: A cube with an edge of 4 cm can contain 8 cubes with an edge of 2 cm. Thus, I have to multiply the weight by 8. So, I do 6 grams × 8 = 48 grams.
>
> Who is right, Wim or Steve?

Wim expressed the dominant misconception that the weight (volume) and the edge of a cube are directly proportional (National Council of Teachers of Mathematics, 1989), whereas Steve argued correctly that the weight of a cube is multiplied by factor 8 when its edge is doubled. Students were not only asked to choose the correct reasoning but also to justify their choice. This confrontation with two alternative solutions and with their underlying reasoning processes was expected to provoke a cognitive conflict. As derived principally from the work of Piaget and his followers, a cognitive conflict arises when new experiences or opinions are perceived by the learner as being in contradiction with his or her existing understanding. Such contradictions and the disequilibrating effect they have on the learner can lead the learner to question his or her beliefs and to try out new ideas (Forman & Cazden, 1985). In Piaget's words,

> [disequilibrating] forces the subject to go beyond his current state and strike out in new directions (1985, p. 10).

Cognitive conflicts, often created in a context of social interaction with the child's peers, drive the individual learner to higher levels of understanding and reasoning (Perret-Clermont, 1980).

In the M–V+ group (the no metacognitive scaffold and visual scaffold group), every item in the test came with an appropriate drawing of the problem situation on squared paper, instead of the plain paper used in the studies reported in chapter 2. An example of such a ready-made drawing is given in Figure 3-1 (belonging to the non-proportional item about squares given in Table 2-1).

Finally, in the M+V+ group (the metacognitive and visual scaffold group), both kinds of help were combined.

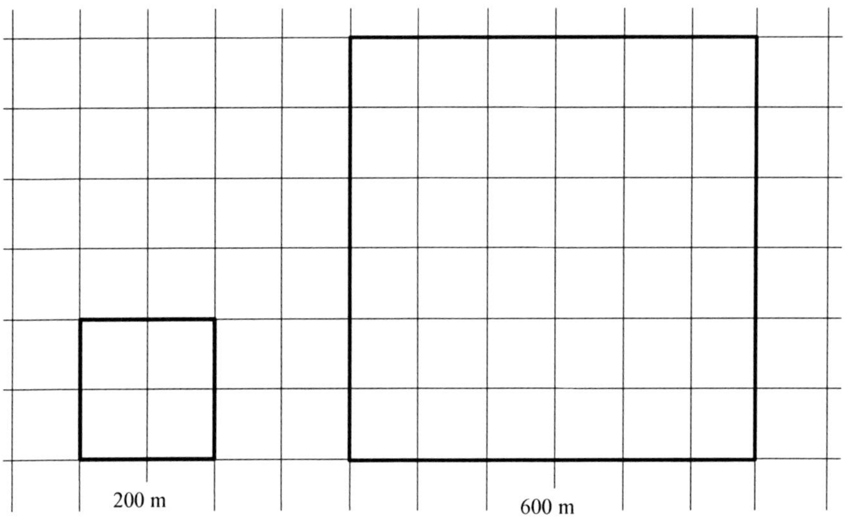

Figure 3-1. Example of a ready-made drawing on squared paper (belonging to the item "Farmer Carl needs approximately 8 hours to fertilise a square piece of land with a side of 200 m. How many hours would he need to fertilise a square piece of land with a side of 600 m?", see Table 2-1.)

Support for the expectation that presenting drawings on squared paper can help students to solve area problems correctly can be found in research findings in the field of cognitive psychology. Mullet, Lautrey, and Glaser (1989) asked 5-year old children to judge areas of rectangles that were presented in four different ways: (1) as blank figures with no reference points at all, (2) with scaled marks on the perimeter, (3) with rows of plus signs on the inside, and (4) paved with small squares. The first two presentations, in which the surface of the rectangle was left blank, more often elicited the well-known erroneous additive pattern for judging area (area = height + width; Anderson & Cuneo, 1978), whereas the last two presentations, in which the rectangles contained graphical features on the inside, increased the number of correct, multiplicative area judgements. Mullet et al. described children's correct judging process as 'visual scanning' of the area. In the mathematics education literature, the expectation that squared paper would be more helpful for the students in the experiment is supported by Clarkson (1989), who observed several 10–13-year old students being able to draw enlargements of geometrical figures on squared paper but unable to perform these drawings on plain paper.

2.2 Hypotheses

Complementary to the hypotheses from the studies which were reported in the previous chapter about (1) the predominance of the linear model, (2) the influence of the age of the students, and (3) the influence of the type of geometrical figure involved, the following hypotheses were stated.

With respect to our first experimental manipulation, we hypothesised that confronting students with a non-proportional problem and forcing them to make a deliberate choice between the incorrect (linear) and the correct (non-linear) solution at the onset of the test, would have a beneficial effect on the mindfulness with which they approach (the non-proportional items in) the test. Therefore, we predicted that the two groups receiving the metacognitive scaffold (M+V– and M+V+) would perform better on (the non-proportional items in) the test than those who did not receive this scaffold (M–V– and M–V+). Second, we hypothesised that the metacognitive support would be more effective for (non-proportional) items about squares than for those dealing with circles or irregular figures. The introduction problem dealt with the weight (volume) of an enlarged cube and proposed a non-proportional solution strategy as one of the two candidate solution methods – namely, filling up the big cube with little ones – which is rather easily transferable to the solution of the (non-proportional) S items. On the other hand, the transfer of this proposed solution strategy to the non-proportional C and I items is much more distant. Third, we hypothesised that the metacognitive scaffold would be more effective in the older age group because these students have greater mastery of the mathematical knowledge and skills needed for solving non-proportional items.

With respect to the second experimental manipulation, we anticipated a better performance on (the non-proportional items in) the test for the two groups who received the visual scaffold (M–V+ and M+V+) than for those who did not receive it (M–V– and M+V–). This prediction was based on the hypothesis that the availability of these drawings on squared paper – in contrast with the drawings on plain paper we used in the studies reported in chapter 2 – would be of considerable help to students who had difficulties in modelling and solving the (non-proportional) items. Placing the drawings on squared paper would help the students see the relations between the lengths and the areas of the two plane figures involved in the problem and would suggest an informal but efficient solution method based on paving. Moreover, we hypothesised that – like the metacognitive scaffold – the visual scaffold would interact with the two other experimental variables, namely, the nature of the figures involved in the (non-proportional) items and the age of the participants. With respect to type of figure, it is clear that providing a drawing of the problem on squared paper is much more helpful when the (non-proportional) item deals with squares than when it involves circles or irregular figures. Therefore, we predicted that the drawing effect would be greater for (non-proportional) S items than for (non-proportional) C and I problems. We further anticipated a greater facilitating effect of the visual scaffold in the older age group because 15–16-year old students are already more experienced in effectively applying heuristic methods (including the use of drawings and other problem visualisations).

Finally, we hypothesised an additive effect of the metacognitive and the visual scaffold. The rationale is that students receiving both scaffolds not only receive a strong warning that not all items in the test are standard proportional problems but are also armed with an extra tool (namely, drawings on squared paper) for modelling and solving these difficult and unfamiliar problems in an intuitive, context-bound, graphical way requiring little or no sophisticated formal mathematical knowledge. Therefore, we predicted that the best performance on (the non-proportional items in) the test would come from the group receiving both the warning and the drawings (M+V+).

2.3 Analysis and results

Similarly as in the studies reported in chapter 2, we categorised all responses on the proportional and the non-proportional items as 'correct' or 'incorrect', not taking into account purely technical mistakes. As in these previous studies, we did not split up the incorrect answers any further in the analysis because they were mostly due to inappropriate proportional reasoning.[10]

All hypotheses were tested by means of a '2 × 2 × 3 × 2 × 2' ANOVA with 'Proportionality' (proportional vs. non-proportional items), 'Age' (12–13- vs. 15–16-year olds), 'Figure' (squares, circles, and irregular figures),

[10] Remember that for the first study reported in chapter 2, we analysed in detail a random sample of 300 incorrect answers on non-proportional items. This analysis revealed that 95% of them resulted from an inappropriate proportional reasoning process.

'Warning' (groups with vs. without metacognitive scaffold), and 'Drawing' (groups with vs. without visual scaffold) as independent variables and the number of 'Correct answers' as the dependent variable. Table 3-1 gives an overview of the percentage of correct responses for the four groups of 12–13- and 15–16-year olds on the proportional and the non-proportional items about squares (S), circles (C) and irregular figures (I) in the test.

Table 3-1. Percentage of correct responses of the 12–13-year and 15–16-year olds on the different categories of proportional and non-proportional items in the test from Follow-up study 1

	12–13-year olds						15–16-year olds					
	Proportional items			Non-proportional items			Proportional items			Non-proportional items		
Group	S	C	I	S	C	I	S	C	I	S	C	I
M-V-	92	98	95	8	5	1	90	94	98	32	15	2
M+V-	92	100	95	9	3	0	74	80	97	44	21	12
M-V+	95	98	90	19	8	2	82	87	97	29	18	3
M+V+	84	90	91	24	10	0	87	78	97	60	23	8

As in the two studies reported in chapter 2, there are very pronounced differences in the performance on the proportional and the non-proportional items (percentages of correct responses were 91% and 15%, respectively). The difference between the two age groups was also observed again. The 15–16-year olds performed better on the test in general than the 12–13-year olds – percentages of correct answers were, respectively, 55% and 50%. The 15–16-year olds performed better than the 12–13-year olds on the non-proportional items (22% and 7% correct responses, respectively), but these better scores on the non-proportional items were accompanied by lower scores on the proportional items (88% and 93% correct responses).

As in the previous studies, the type of figure again had an impact on the number of correct responses. Global percentages for S, C, and I items were in the same direction (58%, 52%, and 49% correct answers, respectively) as in that previous study. Again, this was completely due to the responses on the non-proportional items (28%, 13%, and 4% for the S, C, and I items, respectively), as the percentages of correct responses on the proportional items were in the opposite direction (87%, 91%, and 95% for the S, C, and I items, respectively).

With respect to the first experimental manipulation, i.e., the metacognitive scaffold, we did not find an overall beneficial effect on students' performance on the test as a whole – the overall performance of the students from the M+V– and the M+V+ groups was almost the same as that of the students from the M–V– and the M–V+ groups (53% and 52%, respectively). However, as expected, the warning did have a positive effect on the scores of the non-proportional items[11], but the size of this effect was rather small: The percentage of correct responses in the groups with and without the metacognitive scaffold were 18% and 12%, respectively. The better performance on the non-proportional items in the two M+ groups was paralleled with a weaker score on the proportional items (i.e., 89% correct answers vs. 93% for the M– groups). Furthermore, the hypothesis concerning the differential impact of the metacognitive scaffold on the distinct types of non-proportional problems was confirmed[12]. The percentages of correct answers on the non-proportional S, C, and I items were 34%, 14%, and 5%, respectively, in the groups with warning and 22%, 11%, and 2%, respectively, in the groups without warning, with only the difference for the S items being significant. Therefore, the warning did matter for the S items, but this warning effect did not transfer from the S to the C and I items. Finally, the anticipated greater effect of the warning on students' performance on the non-proportional items in the older age group was confirmed too[13]. The percentage of correct responses on the non-proportional items in the groups with and without warning were 8% and 7%, respectively, for the 12–13-year olds and 28% and 16%, respectively, for the 15–16-year olds. In other words, only the 15–16-year olds took advantage of the metacognitive scaffold.

With respect to the second experimental manipulation, i.e. the visual scaffold, we did not find an overall effect either. The test performance of the students who received the drawings on squared paper was not significantly better than that of those who did not receive these drawings. Percentages of correct answers were, respectively, 53% and 52%. The visual scaffold did have a small but significant positive effect on the scores for the non-proportional items[14] (the percentage of correct responses in the two groups with and without drawings were 17% and 13%, respectively), but once again, the better performance on the non-proportional items in the M–V+ and the M+V+ groups was accompanied by a weaker score on the proportional items (90% vs. 92% correct answers for the M–V– and M+V– groups). Finally, the results did not support the hypotheses that the drawings would have the greatest effect on the performance on the non-proportional S items and in the older age group, implying that the better performance on the non-proportional items and the weaker performance on the proportional items due to the drawings manifested themselves equally in the distinct

[11] 'Warning' × 'Proportionality' interaction effect: $F(1,377) = 15.50$, $p < .01$

[12] 'Warning' × 'Proportionality' × 'Type of figure' interaction effect: $F(2,754) = 5.45$, $p < .01$

[13] 'Warning' × 'Proportionality' × 'Age' interaction effect: $F(1,377) = 7.35$, $p < .01$

[14] 'Drawing' × 'Proportionality' interaction effect: $F(1,377) = 6.96$, $p < .01$

problem types (i.e., S, C, and I problems) and for the two age groups of students (i.e., the 12–13- and 15–16-year olds).

Finally, the results did not support the hypothesis concerning an additive effect of the metacognitive and the visual scaffold, either for the test as a whole, or for the non-proportional items alone.

2.4 Additional questions and findings

The preceding ANOVA revealed significant but remarkably small effects of both the metacognitive and visual scaffold on students' performance on the non-proportional items in the test. However, this analysis did not distinguish between students who took advantage of these scaffolds (i.e., students for whom the warning was effective or who effectively made use of the ready-made drawings on squared paper) and those who did not. To obtain a deeper insight into the effects of the two scaffolds, we also compared the performances on the non-proportional items of the students who effectively noticed and used these scaffolds and those who did not.

To examine the effect of a correct solution of the introductory task on the other non-proportional items in the test, we carried out an additional ANOVA on the non-proportional items in the two M+ groups with 'Intro correct' (correct vs. incorrect intro) and 'Figure' (squares, circles, and irregular figures) as independent variables and the number of 'Correct answers' as the dependent variable.

We hypothesised that the students who solved the introductory task correctly and who thus were obviously receptive to the warning, would perform better on the non-proportional items in the test. Moreover, we hypothesised that this effect of having solved the introductory task correctly would be more powerful for the S than for the C and I items (see above).

The ANOVA revealed a significant main effect of the 'Intro correct' variable. Students who solved the introductory task correctly (i.e., 61% of the 12–13-year olds and 73% of the 15–16-year olds) solved 20% of the non-proportional test items correctly, whereas students who solved it incorrectly solved only 5% of the non-proportional test items correctly. Moreover, the ANOVA revealed a significant interaction between 'Intro correct' and 'Figure'. Students who solved the intro correctly answered, respectively, 38%, 16%, and 5% of the non-proportional S, C, and I items correctly. For the students who did not solve the intro correctly, these percentages were, respectively, 10%, 4%, and 0%.

With respect to the visual scaffold, we were interested in the relation between actually making use of the given drawing on one hand and giving a correct answer on a non-proportional item on the other hand. However, because the notes of the students in the V+ groups did not always provide direct and definitive evidence about the effective use of the given drawing, we had to fall back on an appropriate criterion. Therefore, we distinguished between the drawings that were edited by the student and those that were not. Any student-generated note on the given drawing – even a very small or rudimentary one – was considered an edit. After we had scored all solutions in the two V+ groups in this way, a contingency table was made with the 'Edited' and 'Answer' variables (Table 3-2), and then the (non)independence of the two variables was tested by means of a chi-square test.

Table 3-2. Contingency table for the two groups receiving the visual scaffold (Frequencies and percentages between brackets in Table 3-2 represent the expected cell frequencies and percentages under the null hypothesis of independence of both variables)

	Drawing edited	Drawing not edited	Row totals
Correct answer	48 (17)	121 (152)	169
	4% (1%)	11% (13%)	15%
Incorrect answer	68 (99)	909 (878)	977
	6% (9%)	79% (77%)	85%
Column totals	116	1 030	1146
	10%	90%	100%

The chi-square test revealed that this null hypothesis had to be rejected[15]. A comparison between the observed and expected frequencies in the various cells shows that the dependence of the two variables is in the expected direction: The students who edited the drawings gave more correct answers than would be expected under the null hypothesis, and the students who made no edits on the drawings gave more incorrect answers than expected. Therefore, the actual use of a given drawing on squared paper increases the student's chances of discovering the incorrectness of a stereotypical proportional reasoning with a non-proportional item and, thus, his or her chances of finding the correct solution. On the other hand, this table also shows that this effect was, although significant, not very strong. The effective use of the drawing obviously did not *guarantee* a correct solution.

2.5 Conclusions

In the two studies reported in chapter 2 the strength and omnipresence of the linear model was demonstrated with respect to problems involving length and area of similar plane figures in 12–16-year old students. In the first follow-up study, we investigated the effect of the following two context

[15] $\chi^2(1, N = 1\ 146) = 72.81, p < .01$

variables on the (inappropriate) use of the linear model by students of the same age level who were confronted with the same problem set: (1) the provision of a metacognitive scaffold in the form of an introductory task, which aimed at enhancing the mindfulness with which the students would do the test, and (2) the provision of a visual scaffold in the form of a drawing of each problem made on squared paper, which provided students with useful reference points for measuring lengths and areas and thereby helped them in different ways to model and solve the problems properly.

The major research question was to what extent these two manipulations of the experimental setting would lead to a decrease in the number of incorrect answers based on inappropriate linear modelling on the non-proportional items in the test. The greater the improvement in students' test scores (on the non-proportional items) as a result of these experimental manipulations, the more evidence we would have that the alarming results obtained in the studies reported in chapter 2 were not as frightening as initially thought. However, if we found that the warning and the drawings have only a marginal impact on students' scores (on the non-proportional items), this would yield further and even more convincing support for the strength, the omnipresence, and the obstinacy of the illusion of linearity among secondary school students.

The study yielded significant effects in the expected direction of both kinds of scaffolds on students' performance on the non-proportional items. Offsetting these better results on the non-proportional items in the scaffolded conditions, students' results on the proportional items decreased. Apparently, at least for some students, the scaffolds made it easier to discover and resolve the non-proportional nature of a problem, but as a result, the students sometimes began to question the correctness of the linear model for problem situations in which that model was appropriate. This finding, which is very similar to the one obtained in the studies reported in chapter 2, reveals the fragile and unstable nature of students' emerging non-proportional reasoning scheme. Analogous observations have been made in several other studies about strategic and conceptual change (Siegler & Jenkins, 1989; Vosniadou, 1994, 1999). However, the most important result of this first follow-up study was that the positive effects of the two scaffolds on students' solutions of the non-proportional items were disappointingly small and restricted to certain kinds of problems and age groups. In this respect, we recall that the combination of both scaffolds still yielded 40% incorrect answers on the easiest problem type (S problems) in the older age group (15–16-year olds; see Table 3-1).

3. FOLLOW-UP STUDY 2: EFFECTS OF PROBLEM FORMULATION

3.1 Design and rationale

In the studies reported so far, all proportional and non-proportional items were presented as missing-value problems (e.g., see Kaput & West, 1994; Reiss, Behr, Lesh, & Post, 1985; Tourniaire & Pulos, 1985). In this problem type, three numbers (a, b, and c) are given, and the problem solver is asked to determine an unknown number x. In a proportional missing-value problem, the unknown x is the solution of an equation of the form $a/b = c/x$. Most of the proportional reasoning tasks students encounter in the upper grades of the elementary school and the lower grades of secondary school are formulated in the missing-value format. Textbooks even tend to rely exclusively on this problem format to enhance and assess students' facility with proportional reasoning (Cramer & Post, 1993). At the same time, non-proportional problems stated in a missing-value format are very rare. This practice may contribute to establishing and reinforcing in students' minds a strong association between this problem format and proportionality as a mathematical model. As already discussed in chapter 1, even preservice teachers seem to fall easily into this trap of associating the missing-value format with proportional situations: 32 out of 33 preservice elementary school teachers answered the following non-proportional but simple additive word problem in a proportional way: "Sue and Julie were running equally fast around a track. Sue started first. When she had run 9 laps, Julie had run 3 laps. When Julie completed 15 laps, how many laps had Sue run?" Cramer et al. (1993) commented:

> While these students knew a procedure for solving a proportion, they did not realise that this particular problem did not represent a proportional situation. Therefore, the traditional proportional algorithm was not an appropriate strategy to use (p. 159).

In line with these findings, it could be argued that students' extremely weak results on the non-proportional items in the studies reported so far were not due to intrinsic difficulties with the mathematical concept involved in these problems – namely, understanding the effect of a linear enlargement on area – but mainly the result of a stereotyped problem formulation evoking an unmindful response.

To find out to what extent students' weak performances on the non-proportional items are influenced by this formulation issue, we set up a second follow-up study in which the formulation of the problems was experimentally manipulated while keeping the intrinsic conceptual difficulties constant.

3.2 Method

Table 3-3. Examples of experimental items presented as missing-value and comparison problems

Missing-value problems	Comparison problems
Proportional item about regular figures	
Farmer Gus needs approximately 4 days to dig a ditch around a square pasture with a side of 100 m. How many days would he need to dig a ditch around a square pasture with a side of 300 m? *(Answer: 12 days)*	Farmer Gus dug a ditch around his square pasture. Next month, he has to dig a ditch around a square pasture with a side three times as big. How much more time would he approximately need to dig this ditch? *(Answer: three times more)*
Non-proportional item about regular figures	
Farmer Carl needs approximately 8 hours to fertilise a square piece of land with a side of 200 m. How many hours would he need to fertilise a square piece of land with a side of 600 m? *(Answer: 72 hours)*	Farmer Carl fertilised a square piece of land. Tomorrow, he has to fertilise a square piece of land with a side being times as big. How much more time would he approximately need to fertilise this piece of land? *(Answer: nine times more)*
Proportional item about irregular figures	
On a map of Belgium in an atlas the distance from Genk to Leuven is approximately 5 cm and the distance from Genk to Ghent approximately 11 cm. On a map in front of the classroom the distance from Genk to Leuven is approximately 20 cm. How long is the distance from Genk to Ghent on this map? *(Answer: 44 cm)*	On a map of Belgium in his atlas, Peter measures the distance from Genk to Leuven and the distance from Genk to Ghent. On a map in front of the classroom, the distance from Genk to Leuven is four times as large as in his atlas. How much larger is the distance from Genk to Ghent on this map? *(Answer: four times as large)*
Non-proportional item about irregular figures	
On a map of Belgium in an atlas the distance from Genk to Tongeren is approximately 2 cm and the area of Belgium approximately 250 cm^2. On a map in front of the classroom, the distance from Genk to Tongeren is approximately 6 cm. How large is the area of Belgium on this map? *(Answer: 2 250 cm^2)*	On a map of Belgium in his atlas, John measures the distance from Genk to Tongeren. On a map in front of the classroom, the distance from Genk to Tongeren is three times as large as in his atlas. How much larger is the area of Belgium on this map? *(Answer: nine times as large)*

One hundred and sixty-four 12–13-year old students and 151 15–16-year old students participated in the study. All participants were administered the same paper-and-pencil test consisting of 12 experimental items about the relations among the lengths, areas, and volumes of similar figures, and 3 buffer items. The experimental items belonged to two categories of geometrical figures: 6 items about regular figures (squares and cubes; R items) and 6 items about irregular figures (I items). Within each category of figures, there were 2 proportional items and 4 non-proportional items.

3.3 Hypotheses

The first three hypotheses of this second follow-up study regarding, respectively, (1) the predominance of the linear model, (2) the influence of the age of the students, and (3) the influence of the type of geometrical figure involved, were the same as in Follow-up study 1. Additionally, the following hypotheses were stated.

First of all, we anticipated that rephrasing the problems as comparison problems would have a beneficial effect on students' performance on (the non-proportional items in) the test. When a non-proportional item is formulated as a familiar missing-value problem, it is more likely that students will associate it with the kind of proportional thinking that proved to be the adequate solution strategy for the vast majority of the missing-value problems they had encountered in their school career so far. With the unusual formulation as a comparison problem there is a greater chance that students will no longer fall back on this kind of routine-based, superficial thinking and invest more mental effort in the analysis of the problems, leading to better performance. Besides, in this comparison condition, each problem is given with only one number so that students cannot take refuge in the execution of all sorts of arithmetical operations. In brief, they have no choice other than attending to the mathematical structure of the problem. In line with all previous studies in which better performances on the non-proportional items in a given experimental condition were always paralleled with weaker performances on the proportional items, we hypothesised that the students in the comparison condition would perform worse on the proportional items than the students in the missing-value condition. Accordingly, we predicted for both age groups higher scores on the test for the CO than for the MI group and that this better result would be due completely to the scores of the CO group on the non-proportional items. Finally, we expected that – just as for the two scaffolds from Follow-up study 1 – the formulation of the problems would interact with the two other experimental variables, namely the nature of the figure involved and the age of the students.

3.4 Analysis and results

The hypotheses were tested by means of a '2 × 2 × 2 × 2' ANOVA with 'Proportionality' (proportional vs. non-proportional items), 'Age' (12–13- vs. 15 –16-year olds), 'Figure' (regular vs. irregular figures), and 'Group'

(MI group vs. CO group) as independent variables and the number of 'Correct answers' as the dependent variable.

Table 3-4 gives an overview of the percentages of correct responses for the two groups of 12–13- and 15–16-year olds on the proportional and the non-proportional items about regular figures (R) and irregular figures (I) in the test.

Table 3-4. Percentage of correct responses of the 12–13-year and 15–16-year olds on the different categories of proportional and non-proportional items in the test from Follow-up study 2

	12–13-year olds				15–16-year olds			
	Proportional items		Non-proportional items		Proportional items		Non-proportional items	
	R	I	R	I	R	I	R	I
MI	83	91	22	2	77	96	57	12
CO	49	75	45	17	57	93	68	35

The results indicated, once again, a large difference in the number of correct responses to the proportional and the non-proportional items. For all groups together, the percentages were 78% and 32%, respectively. The 15–16 year olds performed better on all experimental items than the 12–13-year olds (62% and 48% correct answers, respectively), and this difference was more pronounced for the non-proportional items (43% and 22% correct responses, respectively) than for the proportional items (81% and 75% of correct responses). Also the type of figure involved in the problem again had a significant impact on students' performance, with 57% and 53% correct responses on the R and I items, respectively. As in the previous studies, the students obtained higher scores for the non-proportional R items than for the non-proportional I items (48% and 16%, respectively), but their scores on the proportional R and I items were in the opposite direction (66% and 89%, respectively).

With respect to our main experimental manipulation, there was no general effect of the experimental condition on students' performance. In both experimental conditions, the overall percentage of correct responses on the test as a whole was exactly the same (55%). However, the MI group performed considerably worse than the CO group on the non-proportional items (23% and 41% correct answers, respectively), but this worse performance of the MI group on the non-proportional items was accompanied by much better scores on the proportional items (i.e., 87% vs. 68% correct answers in the CO group)[16]. Apparently, the formulation of the

[16] 'Group' × 'Proportionality' interaction effect: $F(1,311) = 118.16, p < .01$

items used in the CO group prevented students from falling into the proportionality trap, but as a result, these students sometimes began to question the correctness of the proportional model for problem situations in which that model was appropriate, a finding that is very similar to the one obtained in all previous studies. Interestingly, the compensation effect was so strong that overall, both groups performed equally well on the test.

3.5 Conclusions

In the previously reported investigations the omnipresence, strength, and obstinacy of the overuse of linearity with respect to school word problems about the effect of a linear enlargement or reduction of a geometrical figure on the perimeter, area, or volume of that figure were convincingly demonstrated. In these studies, the majority of the students failed on the non-proportional items because they routinely applied proportional reasoning in a situation wherein it was inappropriate.

Follow-up study 2 represented a next step in our effort to unravel the factors fostering the occurrence of this illusion of linearity. Typically, the overuse of linearity is qualified as a misbelief (National Council of Teachers of Mathematics, 1989). Students apply the linear model because they incorrectly believe this model is appropriate for a given problem situation. This study demonstrated that students' tendency to apply proportional reasoning in problem situations for which it is not suited, is, at least partly, caused by particularities of the problem formulation, which students had learnt to associate with proportional reasoning throughout their school career. The significantly better results of the CO group on the non-proportional items made it clear that, for a lot of students, what lured them into the trap of proportional reasoning was not their confidence in an overused mathematical model as such (in this case, the linear function) but rather a gradually and implicitly developed association between that model and a particular kind of problem formulation (in this case, the missing-value type). Indeed, the results of the second follow-up study revealed that the missing-value format is an important explanatory factor for the occurrence and persistence of the illusion of linearity in secondary school students. Rephrasing the usual missing-value problems into comparison problems proved to be a substantial help for many students to overcome this illusion. However, remember that students still erroneously solved the rephrased non-proportional items in more than half of the cases!

4. FOLLOW-UP STUDY 3: EFFECTS OF AUTHENTIC CONTEXTS AND DRAWING ACTIVITY

4.1 Rationale

The third follow-up study focused on two other experimental manipulations that were expected to result in a considerable increase of students' success rates on representing and solving word problems about the effect of a linear enlargement or reduction of a geometrical figure on the perimeter, area, or volume of that figure. First, it can be argued that students' weak performance on the non-proportional items in the previous studies is – at least partially – caused by the inauthenticity of the problem context or setting. Over the past decades, several researchers and educators have pointed out the benefits of constructing and organising mathematical activities around rich, attractive and realistic contexts (Cooper & Harries, 2002, 2003; de Lange, 1987; Freudenthal, 1983; Palm, 2002; Treffers, 1987). A dual function is assigned to these contexts: (1) They act upon students' intrinsic motivation and task involvement and thus enhance the likelihood that they will make a serious effort to complete the problem, and (2) they help them to make a correct representation of the problem and to find a correct solution strategy by eliciting the activation and use of prior contextualized knowledge (real-world experiences, intuitions, models, strategies, …) that can be helpful for understanding and solving the problem. According to the above-mentioned authors, realistic contexts may refer not only to aspects of the 'real' social or physical world; they may also refer to imaginary, fairy-like worlds as long as they are meaningful, familiar, and appealing to the students. It is not the amount of realism in the literal sense that is crucial for considering a context as realistic or authentic, but rather the extent to which it succeeds in getting students involved in the problem and engages them in situationally meaningful thinking and interaction. The problems used in all previously reported research were traditional word problems built around rather 'poor' contexts having no special meaning or attractivity to 12–16-year youngsters nowadays. Therefore, students' performance on the non-proportional items might increase significantly if we could succeed in increasing the authenticity of these problems for the students. Some support for this assertion can be found in the work of scholars of the Freudenthal Institute, who have explored students' difficulties with the influence of linear enlargement on area and volume in realistic contexts (like the instructional units 'With the giant's regards' in Streefland, 1984, or 'Gulliver' in Treffers, 1987). In 'With the giant's regards' for example, the hand of a giant, being four times as long as a human hand, was drawn on the blackboard and pupils of Grade 3 were asked to estimate various lengths, areas, and volumes in the giant's world (e.g. How big would the giant's newspaper be? How much would he weigh? …). According to Streefland (1984), these activities proved to be helpful for many students in breaking the 'pattern of linearity' and making them aware of the 'multi-dimensional' impact of increase on area and volume. Treffers (1987) adopted and elaborated these ideas in a small-scale classroom study with 6th-graders built around the context of 'Gulliver's travels'. In his report, Treffers (1987, p. 5) states:

> This question [*How many Lilliputian handkerchiefs make one for Gulliver?*] introduces the influence of linear enlargement on area. The students have no difficulty with the problem.

Unfortunately, Treffers does not support this latter claim with empirical data.

Second, there is the vast amount of theoretical and empirical research on how and why drawings and diagrams may enhance people's ability to represent and solve mathematical problems (e.g., Aprea & Ebner, 1999; De Corte et al.,1996; Hall et al., 1997; Larkin & Simon, 1987; Pólya, 1945; Reed, 1999; Schoenfeld, 1992; Vlahovic-Stetic, 1999). Self-made drawings, in particular, may stimulate a more deep-level and mindful approach to a task (Aprea & Ebner, 1999; Dirkes, 1991). The construction of a drawing induces a systematic analysis and elaboration of the problem situation, it enhances a planned solution of the task, and the drawing can be used in interpreting and checking the answer. We need to keep in mind, however, that the two studies reported in chapter 2 showed no significant positive effect either of the instruction to make drawings or of providing ready-made drawings. In Follow-up study 1 (reported earlier in the present chapter) the positive effect of offering drawings on squared paper was statistically significant but still disappointingly small. In retrospect, the absence of clear-cut positive effects of drawings in all those studies might have been caused by the following elements: (1) Only a very small number of students in those studies effectively made a drawing themselves or actually used the given drawing, and (2) due to the way the drawings and the drawing instructions were presented in those studies, the drawings did not constitute a genuine part of the problem-solving process. For these reasons, we were reluctant to consider the absence of a positive effect of drawings or drawing instructions in those studies as convincing evidence against the drawing hypothesis, but rather as an encouragement to operationalise the drawing factor in a more appropriate way. We therefore hypothesized that when the drawing activity becomes a truly integrative part of students' problem-solving process, and when students' drawings effectively show the topological and geometrical relations among the essential components of the problem, a significant improvement of their scores on non-proportional items will be obtained.

4.2 Design

One hundred and fifty-two 13–14-year olds and 161 15–16-year olds participated in the study. The participants came from a school for general secondary education located in a medium-sized Flemish town and pursuing high educational standards. Boys and girls were almost equally represented in the school, and also in our sample. All participants were confronted with a paper-and-pencil test consisting of six experimental items (two proportional items and four non-proportional items) about the relationships among the lengths, areas, and volumes of different types of rectilinear and non-rectilinear figures, together with some buffer items. To allow a proper evaluation of the two experimental manipulations mentioned above, we applied a 2 × 2 design with four matched groups at both age levels. The matching of the four groups was done by guaranteeing that (1) the different study streams provided in the school were equally represented in the four groups, (2) the average number of hours a week students from the four groups spent at mathematics was the same, and (3) students' average result on the previous year's final mathematics examination did not significantly differ between the four groups (for more details about this matching procedure, see Claes, 2000). Four different versions of the test were administered: Problems were presented in or out of an *a*uthentic context or setting (respectively, the A+ and A– condition), and either with or without an integrated *d*rawing instruction (respectively, the D+ and D– condition), leading to four combinations (A+D+, A+D–, A–D+ and A–D–) that were administered to the four matched groups at each age level.

In the two A+ groups, students' involvement in the problems was experimentally enhanced by preceding the test by a series of well-chosen video fragments of a screen version (Kenworthy & Sturridge, 1996) of Jonathan Swift's world-famous novel 'Gulliver's travels', first published in 1726. All fragments were selected from the story of Gulliver's visit to the isle of the Lilliputians, a world in which all lengths are 12 times as small as in our (and Gulliver's) world, and were assembled to tell in ten minutes the whole plot in a visually attractive way. To direct students' attention to the relevant features while watching these video fragments, they were instructed to look carefully at the ratios between the Lilliputian and Gulliver's world and it was announced that they would have to solve a test consisting of mathematical problems linked to the video fragments afterwards. Because the linear scale factor was imposed by the context, all items involved reductions by factor 12, which was clearly announced in the following way at the beginning of the test.

> In the world of the Lilliputians, all lengths are 12 times as small as in our world, the world of Gulliver. A tree, being 12 m high in our world, would thus be only 1 m high in the world of the Lilliputians. A road, being 12 km long in our world, would be only 1 km long in the Lilliputian world. And, of course, the Lilliputians themselves measure also 12 times smaller than Gulliver.

Table 3-5. Experimental items used in the A+ and A– groups

Test items in A+ groups	Test items in A- groups
Proportional items	
Gulliver's walking-stick is 96 cm high. What's the height of a similar Lilliputian walking-stick? *(Answer: 8 cm)*	The length of a line segment A is 13 times as large as the length of a line segment B. If the length of line segment A is 78 cm, how long is line segment B? *(Answer: 6 cm)*
Gulliver's waistband is 108 cm long. What's the length of a similar Lilliputian waistband? *(Answer: 9 cm)*	The length of a line segment C is 11 times as large as the length of a line segment D. If the length of line segment C is 132 cm, how long is line segment D? *(Answer: 12 cm)*
Non-proportional items	
Gulliver's handkerchief has an area of 1 296 cm^2. What's the area of a similar Lilliputian handkerchief? *(Answer: 9 cm^2)*	The side of square Q is 12 times as large as the side of square R. If the area of square Q is 1 440 cm^2, what's the area of square R? *(Answer: 10 cm^2)*
The sole of Gulliver's shoe has an area of 288 cm^2. What's the area of the sole of a similar Lilliputian shoe? *(Answer: 2 cm^2)*	The diameter of a circle E is 11 times as large as the diameter of a circle F. If the area of circle E is 242 cm^2, what's the area of circle F? *(Answer: 2 cm^2)*
In Gulliver's world, a cheese cube has a volume of 1 728 mm^3. What's the volume of a Lilliputian cheese cube? *(Answer: 1 mm^3)*	The side of a cube I is 13 times as large as the side of a cube J. If the volume of cube I is 2 197 cm^3, what's the volume of cube J? *(Answer: 1 cm^3)*
In Gulliver's world, a wineglass has a volume of 1 728 mm^3. What's the volume of a similar wineglass in the Lilliputian world? *(Answer: 1 cm^3)*	The diameter of sphere M is 12 times as large as the diameter of a sphere N. If the volume of sphere M is 172 800 mm^3, what's the volume of sphere N? *(Answer: 100 mm^3)*

All problems were stated in a so-called comparison format, in which one measure (a length, area, or volume) of an object in Gulliver's world was given, and students had to determine a corresponding measure in the Lilliputian world (which, arithmetically, leads to divisions by 12, 12^2, or 12^3

for, respectively, one, two, or three dimensional magnitudes). In the two corresponding A– groups, the experimental items were equivalent from a mathematical point of view, but were presented in the form of a series of non-related traditional school problems without any attempt to embed the mathematical task in a contextually meaningful and attractive setting. To bring some variation in the test, three different reducing factors were used (11, 12, and 13). Table 3-5 gives an overview of the test items used in the A+ and A– groups.

A+ groups

Below the handkerchief of Gulliver is drawn. Draw a similar handkerchief of a Lilliputian beside it.

Gulliver's handkerchief *Handkerchief of a Lilliputian*

A– groups

The side of square Q is 12 times as large as the side of square R. Below square Q is drawn. Draw square R beside it.

Square Q *Square R*

Figure 3-2. Example of presentation of drawing activity in A+ and A– groups

In the D+ groups, students were provided with a drawing of the (geometrical) object introduced in the problem and they were first asked to complete the drawing by making, next to the given original, a reduced copy using the given scale factor (which was, as mentioned before, always 12 in the A+ groups, and 11, 12 or 13 in the A– groups). To assure that the students would perform the drawing task as a genuine part of their solution process, the drawing instruction was given before all necessary numerical data were presented. Figure 3-2 exemplifies how this drawing task was presented on students' response sheets for the first non-proportional item in the A+ and A– groups given in Table 3-5.

In all four experimental conditions, students could not only write the answer on their response sheets, but also make calculations, sketches, or comments. When they had finished the test, students were asked to complete a short questionnaire consisting of several statements which had to be rated on a five-point scale (from full agreement to full disagreement). Beside six general statements (G1, G2, G3, G4, G5, and G6) submitted to all participants, there were two *v*ideo related statements (V1 and V2, of course, only submitted in the A+ groups), three *d*rawing related statements (D1, D2, and D3, for the D+ groups) and one *v*ideo *and d*rawing related statement (VD1, for the A+D+ group). Table 3-6 gives an overview of these statements. The main purpose of the questionnaire was to provide some information on how the test and the experimental setting were perceived and interpreted by the students.

4.3 Hypotheses

Additionally to some general hypotheses which already were formulated and tested in the previous studies, the following hypotheses were stated.

First, based on the claims of the advocates of authentic or realistic mathematics education (see above), we hypothesised that plunging students into a rich and attractive context – in our case, the story of Gulliver's visit to the Lilliputians, coming alive in a fantastic video faithfully respecting the original 1:12 ratio for lengths – would help them to engage themselves in meaningful and thoughtful mathematical thinking and problem solving, especially on the difficult non-proportional items. Accordingly, we predicted for both age groups higher scores for the A+ groups than for the A– groups, due to a positive effect of the authenticity factor on students' willingness and capacity to model and solve the non-proportional items.

Second, we anticipated a better performance, especially on the non-proportional items, for the students who had to make a drawing of a figure similar to a given original before the start of their solution process. This hypothesis is in line with numerous investigations reporting positive effects of drawings and diagrams, and especially of self-made visualisations, on students' mathematical problem solving (see above). Accordingly, we predicted for both age groups higher scores for the D+ groups than for the D– groups, due to the anticipated positive effect of making a drawing on students' capacity to represent and solve the non-proportional items.

Table 3-6. Questionnaire used in the different experimental conditions

General statements
(G1) I think I shall have a good score on this test.
(G2) I liked to work on this test.
(G3) The problems are similar to the problems we solve in the classroom.
(G4) I did my best to work on this test as much as possible.
(G5) I considered this as an easy test.
(G6) It as an instructive experience to complete this test.
Video statements
(V1) While solving the arithmetical problems, I often recalled the video fragments on Gulliver.
(V2) The video fragments helped me to find the answer.
Drawing statements
(D1) While solving the arithmetical problems, I made use of the drawing above.
(D2) The drawings helped me to find the answer.
(D3) I usually make a drawing before solving a geometrical problem.
Video and drawing statement
(VD1) The video fragments helped me to make the drawing.

4.4 Analysis

In view of the quantitative data analysis, all responses on the proportional and the non-proportional items were categorised as 'correct' or 'incorrect'. A response was considered as 'correct' when it was the result of a mathematically appropriate reasoning process. Accordingly, answers that differed from the correct answer because of a purely technical mistake in a correct overall solution process were considered as 'correct' too. All other kinds of erroneous answers were scored as 'incorrect'.

The hypotheses were tested by means of a '2 × 2 × 2 × 2' ANOVA with as independent variables the within subject variable 'Proportionality' (proportional vs. non-proportional items) and the between subject variables 'Age' (13–14- vs. 15–16- year olds), 'Authenticity' (A+ vs. A– groups) and 'Drawing' (D+ vs. D– groups), and the number of 'Correct answers' as the dependent variable. To obtain the most pure and valid comparison between the A+ and A– groups, we decided to restrict the above-mentioned quantitative analysis to only four of the six experimental items (the two proportional items and the two non-proportional items about rectilinear figures, and thus not the two non-proportional items about non-rectilinear figures)[17]. However, as documented convincingly by Claes (2000), the results of this restricted analysis are representative for the whole data set.

This quantitative analysis, aiming at testing the four above-mentioned hypotheses, was complemented by a qualitative analysis of students' solution strategies and their schematic drawings. Students' answers to the different statements of the questionnaire were analysed for each statement separately.

4.5 Quantitative results

As in the previous studies, the overall percentages of correct responses for the proportional items were much higher than those for the non-proportional items (i.e., 95% and 33%, respectively). The 15–16-year olds performed better than the 13–14-year olds on all experimental items (69% and 59% correct answers, respectively), and this difference was due to the performance on the non-proportional items. The 15–16-year olds answered nearly twice as many non-proportional items correctly than the 13–14-year olds (43% and 23% correct responses, respectively), while the performance of both age groups on the proportional items was exactly the same (95% correct responses).

With respect to the authenticity hypothesis, we found a performance difference between the A+ and A – groups, but the effect was in the opposite direction. Students who watched the video fragments and received the video-related items performed significantly *worse* than the students from the other groups (61% correct responses for the A+ groups versus 67% for the A– groups)[18]. For the proportional items, the results were in the expected direction (respectively 97% and 93% for the A+ and A– groups), but the percentage of correct responses on the non-proportional items in the A+ groups (25%) was much lower than in the A– groups (41%)[19].

The results also did not support the drawing hypothesis. On the contrary, as for the authenticity factor, students who had to make a drawing even

[17] In the A– groups, the non-proportional items about non-rectilinear figures handled about circles and spheres, geometrical forms students typically encounter in school geometry, while in the A+ groups, these forms were replaced by 'real' but different non-rectilinear forms issued from Gulliver's world (a sole of a shoe and a wineglass). For the non-proportional items about rectilinear figures, the same geometrical forms (squares and cubes) could be used in both experimental conditions.

[18] 'Authenticity' main effect: $F(1,302) = 5.91$; $p < .05$

[19] 'Authenticity' × 'Proportionality' interaction effect: $F(1,302) = 12.43$, $p < .01$

performed significantly *worse* than students from the non-drawing groups (59% correct responses in the D+ groups versus 70% for the D– groups)[20]. The negative effect of making a drawing was much greater for the non-proportional items (respectively 23% and 44% correct responses for the D+ and D– groups) than for the proportional items (respectively 94% and 96% correct responses for the D+ and D– groups)[21].

Because of the very surprising nature of the results with respect to the two main hypotheses, we decided to do a replication of this follow-up. Although the matching of the students in the four groups had been done very carefully and systematically taking into account all available possibly relevant information about the students (see Claes, 2000), and although the administration of the test happened as planned, one cannot exclude that the results might have been accidentally affected by some unattended initial differences between the four experimental groups. For practical reasons, we restricted the replication study to only one of the two experimental manipulations. We choose the drawing factor because its negative results had surprised us most.

One hundred and sixty-five students from the same age groups but from a different school and region participated in this replication study. Using all available possibly relevant information (the study stream to which the students belonged, the number of hours a week spent at mathematics, and students' examination results for mathematics), the students were individually matched in a drawing and a no-drawing group, defined in the same way as in the original study. The same paper-and-pencil tests were administered, and we absolutely guaranteed that all participants could work as long as they wanted to complete the task. The results of that replication study were exactly the same, in the sense that the students who were instructed to make drawings again performed significantly worse on the test in general[22] (60% versus 73% correct responses in, respectively the drawing and non-drawing condition) and on the non-proportional items in particular[23] (40% versus 62% correct responses in, respectively the drawing and non-drawing condition) (for a detailed report of this replication study, see Claes, 2000).

[20] 'Drawing' main effect: $F(1,302) = 19.52$; $p < .01$

[21] 'Drawing' × 'Proportionality' interaction effect: $F(1,302) = 10.38$, $p < .01$

[22] 'Drawing' main effect: $F(1,159) = 9.27$; $p < .01$

[23] 'Drawing' × 'Proportionality' interaction effect: $F(1,159) = 22.95$, $p < .01$

4.6 Additional findings

As mentioned above, we also collected qualitative data about the nature of students' solution strategies and their drawings. Besides, we administered a questionnaire consisting of a number of statements about how the students had perceived and appreciated the test. Hereafter we briefly report the major results of these additional analyses, restricting ourselves to those findings that are of direct help to understand the surprising results with respect to the authenticity and the drawing hypothesis. For a more detailed and more systematic overview of the findings from these additional analyses, we refer again to Claes (2000).

Students' solution strategies. A detailed analysis of the written protocols of all students' correct solution strategies for the non-proportional items was undertaken. Because only the non-proportional items dealing with rectilinear figures were included in the quantitative analysis, we also restrict the present discussion to those items. In the studies reported in chapter 2, the solution processes underlying a correct answer on a non-proportional item were categorised using the following classification schema: (1) 'paving': finding the area of a plane figure by paving it with small, similar figures, (2) 'formula': finding and applying an appropriate mathematical formula for the area, (3) 'general principle': applying the general rule 'length $\times r$, thus area $\times r^2$'. For the present study, we had to adapt this former classification in two different ways. First, because the present study included also items dealing with solids, the 'paving' category should also include its three-dimensional variant, finding the volume of a solid by filling it up with small, similar solids. Second, we added a fourth and fifth category: (4) 'mixed' strategy for the responses found by means of a combination of at least two different strategies, and (5) 'remainder' to include the correct solutions for which the solution strategy could not be determined.

Table 3-7 gives an overview of solution strategies students used to correctly solve a non-proportional item about a rectilinear figure in the different experimental conditions.

Altogether, 46% of the correct responses on non-proportional items about rectilinear figures were found by applying the 'formula' strategy. Besides, the analysis revealed a remarkable difference between the A– and A+ groups: In the A– groups many more correct solutions were found by applying an appropriate mathematical formula than in the A+ groups. Apparently, the traditional context more often elicited this school-like approach, which not only prevented the students from falling into the linearity trap but also brought them to the correct solution. The 'general principle' strategy – 'length $\times r$, thus area $\times r^2$' – was the second most popular strategy, accounting for 42% of the correct responses on non-proportional items about rectilinear figures. In the A+ groups, the rare use of the 'formula' strategy was offset by students' application of the 'general principle'. The three other strategies were chosen far less often (for the 'mixed', 'paving' and 'remainder' category, we found percentages of, respectively, 8%, 2%, and 1%). Although in line with students' well-documented resistance to informal, context-bound mathematical strategies (see, e.g. Schoenfeld, 1992; Verschaffel & De Corte, 1997, but also the studies reported in chapter 2), the very rare use of the intuitive 'paving' strategy was surprising. In the D+ groups, not a single student (visibly) found the correct solution to a non-proportional item about a rectilinear (!) figure by paving. Remarkably, the drawing instruction absolutely did not elicit this informal but efficient graphical strategy by the students.

Table 3-7. Number and percentage of strategies underlying students' correct responses on the non-proportional items about rectilinear figures

	Formula		Principle		Mixed		Paving		Remainder	
Group	Number	%	Number	%	Number	%	Number	%	Number	%
A+ (75)[a]	8	11	61	81	2	3	4	5	0	0
A- (131)	87	66	26	20	14	11	1	1	3	2
D+ (72)	30	42	34	47	7	10	0	0	1	1
D- (134)	65	49	53	40	9	7	5	4	2	1
Total (206)	95	46	87	42	16	8	5	2	3	1

Note [a] The numbers between brackets refer to the number of correct responses on the non-proportional items about rectilinear figures for each experimental group.

Analysis of students' drawing activity. Because students' actual drawing activity could also shed more light on the quantitative results of the test, we scrutinised their notes on the presence of drawings of the non-proportional problems, and, in the case a drawing was actually made, we verified its correctness. For a proper interpretation of the unexpected results with respect to the hypothesised effect of the drawing factor on students' performance on the non-proportional items, it was very important to know if the students from both groups (D+ and D–) had actually made drawings for those items and if so, if they had done it correctly. The criteria for scoring *if* a drawing was made were the same in the D+ groups and the D– groups. Each graphical representation referring to the geometrical shape described in the item was scored as a drawing. Mere visualisations or schematisations of the numerical data, such as arrow diagrams, tables or graphs, were thus not considered as drawings. The scoring of *the correctness of* the drawings was done somewhat differently in the D+ and D– conditions. Remember that in the D+ groups, students were explicitly instructed to draw a reduced copy of the geometrical shape described in the item, starting from an original provided on their response sheet and using a given reducing factor (see Figure 3-2). To evaluate the correctness of this drawing, we tolerated a 20% error margin. If the linear dimensions of the students' drawing of the geometrical figure differed 20% or less from its correct linear dimensions, the drawing was scored as a correct one (e.g., if the student had to draw a square with a side of 1 cm, his drawing of that square was considered as a correct one if the side measured between 0.8 and 1.2 cm). In the D– groups, who received no (partial) drawings nor instructions to make a drawing, we additionally required that both the original geometrical form and its reduced copy were drawn. Table 3-8 gives an overview of the number of correct and incorrect drawing in the D+ and D– conditions.

Table 3-8. Numbers and percentages of drawings in the D+ and D– conditions

	Correct drawing		Incorrect drawing		No drawing	
Group	Number	%	Number	%	Number	%
D+ (624)[a]	485	78	99	16	40	6
D- (616)	6	1	58	9	552	90

Note[a] The numbers in brackets refer to the numbers of non-proportional items administered.

The instruction to make a drawing in the D+ groups was followed in 94% (584/624) of the cases. This percentage was remarkably high, compared with the results of the studies reported in chapter 1, in which only 46% of the students followed the *recommendation* to make a drawing *after* the numerical data of the problem were presented. So, from that point of view, the change in the operationalisation of the drawing factor had been very successful. Moreover, 83% (485/584) of these drawings in the D+ groups were 'correct' ones, suggesting that students' numerous errors on the non-proportional items typically did not originate in the phase of the problem representation. In the D– groups, only a minority of the students made visibly effective use of the drawing heuristic: Spontaneously made drawings were found in only 10% (64/616) of the cases and, in addition, only 9% (6/64) of these drawings could be scored as 'correct'.

The results on students' drawings at least made it clear that the significantly lower score of the D+ groups on the non-proportional items could not be attributed to the fact that the students in these groups had neglected the drawing instruction nor to the fact that they had made incorrect drawings. On the contrary, the very high percentage of actual and correct drawings followed by incorrect responses to the non-proportional items in the D+ condition suggests that making a (correct) drawing rather lured them into the proportionality trap instead of protecting them from it – a conclusion that is hard to draw in the light of our current knowledge on the role of heuristics in general, and of visualisations in particular, in skilled (mathematical) problem solving. We come back to this issue in the discussion section.

Results with respect to the questionnaire. Students' degree of agreement with the different statements in the questionnaire (see Table 3-6) was first analysed at a descriptive level by looking at the frequency distributions and by determining students' mean ratings for the different statements in the four different experimental conditions and for all four conditions together. For calculating the mean ratings, we assigned numbers to the values of the scale in the following way: 1 = full disagreement, 2 = disagreement, 3 = no opinion, 4 = agreement, and 5 = full agreement. An overview of the mean ratings in the different experimental conditions and in the total group is given in Table 3-9.

Table 3-9. Mean rating (on a maximum of 5) of the questionnaire statements in the different experimental groups and for the total group

Group	G1	G2	G3	G4	G5	G6	V1	V2	D1	D2	D3	VD1
A+D+	3.24	3.28	1.84	3.93	3.45	3.07	2.34	2.04	3.00	2.86	2.90	2.16
A+D-	3.50	3.61	1.81	4.12	3.55	3.03	2.62	2.01				
A-D+	3.11	3.15	2.11	4.06	3.17	3.19			3.84	3.71	3.34	
A-D-	3.10	3.17	2.30	4.20	3.18	3.16						
Total	3.23	3.30	2.02	4.08	3.33	3.11	2.48	2.03	3.42	3.29	3.12	2.16

Because a comparison between the A+ and A– groups could reveal important information for both the validation of our experimental design and for the interpretation of the results with respect to the authenticity hypothesis, possible differences in students' rating of the general statements (G1, G2, G3, G4, G5, and G6) in both groups were statistically tested by means of a chi-square test. We limit the discussion of the results to the most important findings (for a detailed description, see Claes, 2000).

First, with respect to the six general statements, no clear conclusions could be drawn from students' rating of the statements G1, G2, and G5 in the total group, but the chi-square test revealed some interesting differences between the A+ and A– groups. The A+ groups were more confident to get a good score on the test[24] (statement G1), liked more to work on the test[25] (statement G2) and considered this test as easier[26] (statement G5) than the A– groups. We consider the greater task attractiveness in the A+ groups (statement G2) as support for the validity of our design, because a major purpose of the authentic setting consisted in the creation of an attractive and

[24] $\chi^2(2, N = 303) = 6.91, p < .05$
[25] $\chi^2(2, N = 302) = 11.02, p < .01$
[26] $\chi^2(2, N = 295) = 9.09, p < .05$

stimulating context for the students, and, therefore, in an increase of their involvement in the task. With respect to statement G3, 74% of the students in the total group (fully) *dis*agreed that the problems in the test were comparable with those they had to solve in the classroom. However, students in the A– groups more often agreed with this statement (18%) than those in the A+ groups (6% of the students)[27]. Apparently, the students in the A– groups experienced the presentation and formulation of the problems as more traditional and school-like than those in the A+ groups. With respect to statement G4, 82% of all students declared they did their best working on the test; with no significant difference between the A+ and A– groups. With respect to the last general statement G6 (inquiring if students considered the task to be an instructive experience), the reactions were neutral – the frequency distribution was more or less symmetric around 'no opinion' that itself represented 42% of the data – and no significant differences between the A+ and A– groups were found.

Second, with respect to the two video-related statements (V1 and V2), the general tendency was rather negative: 61% of the students in the A+ groups claimed not to have recalled the video fragments of Gulliver while solving the problems (statement V1) and 77% asserted that these fragments had not helped them to find the answers on the problems (statement V2).

Third, with respect to the three drawing-related statements (D1, D2, and D3), 63% of the students in the D+ groups claimed they had made use of the drawings while solving the problems (statement D1), 55% asserted that the drawings helped them to find the answer (statement D2) and only 44% claimed they usually make a drawing before solving a geometrical problem (statement D3). Apparently, the drawing instruction was perceived as more effective than the presentation of the video.

Finally, with respect to the video *and* drawing related statement VD1, 72% of the students in the A+D+ group claimed that watching the video did not help them to make the drawings.

4.7 Discussion

Previous research convincingly demonstrated that the vast majority of students fail to solve problems involving the relationships between the linear measurements and the area and/or volume of similarly enlarged or reduced geometrical figures, even when considerable help is provided (giving metacognitive support, presenting the problem in another format, …). In the third follow-up study, we tested the effect of two additional forms of help,

[27] $\chi^2(2, N = 302) = 5.22, p < .05$

namely increasing the authenticity of the problem context and asking students to make a drawing as an integral part of the solution process, on students' performance on this type of geometrical problems. Contrary to our expectations, neither of the experimental manipulations yielded the expected result. In fact, they both yielded a significant negative effect on students' performance on the test in general and on the non-proportional items in particular.

With respect to the *absence of a positive authenticity effect* one can point a finger at the way 'authenticity' was operationalised in the present study. Compared to the studies by Streefland (1984) and Treffers (1987) in which students were actively involved in a series of math lessons around an authentic context with several learning tasks (e.g. drawing a giant's footprint, making a newspaper for the giant, ...), watching a video and completing a paper-and-pencil mathematics test about this context seems a rather weak operationalisation of the authenticity factor. Although the video did have a positive effect on students' task-specific motivation (cf. statement G2), the majority of the students claimed that the video fragments had not helped them to find the answers on the test items (statement V2). Probably, a more performance-based form of assessment (involving, for instance, subtasks like effectively 'making a Lilliputian handkerchief' or, 'filling a Lilliputian wineglass') (see also Reusser & Stebler, 1997b) is needed to help students discover the non-linear nature of the problem situations. Later in this monograph (see chapter 6 'Stepping outside the classroom'), we will report an empirical study that was specifically aimed at verifying this hypothetical conclusion.

While the way in which the authenticity factor was operationalised in the present study can account for the absence of a positive effect, it cannot explain the observed *negative* effect. To explain this negative effect, several hypothetical explanations can be raised. Although some of these explanations are (partially) sustained by students' reactions on the questionnaire, we acknowledge their speculative nature. First, the way in which the problems were stated and presented in the A– condition was more similar to how these problems are typically formulated and presented in mathematics lessons at school (some support for this explanation can be found in the results with respect to the questionnaire). A second possible explanation can be found in Salomon's (1981) account of the mediating effects of people's perceptions of media characteristics on their willingness to invest mental effort during learning and problem solving, and, consequently, on their performance. According to Salomon, students perceive video as a less difficult medium than written materials, and therefore are inclined to invest less mental effort in working with information transmitted by this easy medium as compared to media that are perceived as difficult. Once again, some support for this explanation can be found in the questionnaire. Students in the authentic setting perceived the difficulty level of the test lower and their subjective competence higher than students in the traditional setting (see the results with respect to, respectively, the statements G5 an G1). For a third possible explanation, we refer to a related finding reported by Boaler (1994), namely that students – and especially girls – were likely to underachieve in realistic mathematical contexts as compared to problems presented in an abstract way. Boaler analysed the performance of 50 students from a school with a traditional approach to mathematics education on two sets of three questions assessing the same mathematical content through different contexts. Interestingly, the girls from that school attained a lower grade on an item built around the context of fashion than on an isomorphic item that was embedded in an abstract context or in a context that was less appealing for these girls (e.g. football). According to Boaler, these girls' underachievement on the fashion item was caused or influenced by their greater involvement in the problem context, which led them away from its deeper mathematical structure. Although there are quite some differences between the way in which the authenticity factor was operationalised in the two studies, it is possible that students' emotional involvement in the fantastic world of the Lilliputians may have had a negative instead of a positive influence on their performance, just like for the girls in Boaler's study. A fourth and possible explanation refers to the time factor. Students from the A+ groups, who first watched the video fragments for ten minutes, had less time for solving the test than students from the A– groups. Although students from all groups were not forced to give in the test before they were ready, the time factor may have played a more subtle negative role in the A+ groups. Referring to the notion of 'experimental contract' (Greer, 1997), which can be described as a system of implicit norms, rules and expectations that influence how a participant thinks and acts during an experiment, it could be argued that the time provided for a test yields signals about the nature, in particular complexity, of the task(s) to be fulfilled. Consequently, the less time the participants are given for completing a task, the more chance they will perceive it as easy. The finding that the students in the authentic setting perceived the test as easier than those of the traditional setting (statement G5) is in line with this assertion.

The *absence of a positive effect* of asking students to make a drawing both in the major study and in its replication, was not in agreement with our drawing hypothesis, but corresponds with a number of other studies reporting either no or only marginal effects of the use of heuristics (like making a diagram or a drawing) on (mathematical) problem solving (De Corte et al., 1996; Schoenfeld, 1992; Van Essen & Hamaker, 1990). Among the reasons that are given in this literature for the weak relationship between heuristics and success in (mathematical) problem solving are the following: (1) The descriptions of the heuristics are not sufficiently detailed to enable students who are not yet familiar with them to implement and use them efficiently in their problem-solving endeavours; (2) heuristics are only efficient when they are inculcated in conjunction with metacognitive executive control skills. Whereas the first of these explanations seems not very helpful to explain our results (the qualitative analysis of the students' drawings revealed they were mostly correct), the second one seems particularly relevant (see below). But, still, these explanations cannot account for the strong *negative* effect of making a drawing, which has, as far as we know, never before been reported in the research literature on the role of heuristics in general and of visualisation in particular, in (mathematical) problem solving. Having replicated these remarkable results in a follow-up study with the drawing factor as the only experimental variable (as briefly reported above), it became clear that the explanation for this remarkable result had to be sought in the precise role played by the drawing heuristic in students' problem-solving processes. Ultimately, a detailed rational analysis of the cognitive processes of students' solving of the non-proportional items in the D+ conditions yielded an explanation for this remarkable finding, which needs, however, empirical validation in further research. When students are asked to make a reduced copy of the given geometrical shape (cf. Figure 3-1), they first measure a linear element of that shape (e.g. the side of the given square, the height of the wineglass, …) and then divide that element by the given linear scale factor to obtain the length of the corresponding dimension of the reduced shape that is needed for drawing this reduced shape. So, what they have in their working memory at the beginning of their actual solution process of a non-linear problem is exactly the *linear scale factor* they just applied in the drawing activity, rather than its square or cube which is needed for the solution of the problem. The activation of the linear scale factor in students' working memory during the drawing task may have enhanced the (erroneous) inclination to re-use that activated information element in the actual solution of the problem through a kind of automatic priming mechanism. Another mechanism that may underlie this inclination is students' more or less deliberate reliance on the experimental contract (Greer, 1997). If the experimenter involves the students in a preparatory activity, such as the making of a drawing, they will normally expect that this activity will be of help for solving the subsequent problem. Most likely, in our case this preparatory activity was perceived by the students as merely a matter of 'dividing by the linear scale factor', and it was exactly the transfer of this information element to the subsequent solution of the problem that lured them into the linearity trap.

Follow-up study 3 has added two factors to the list of candidate forms of help which have proven to be of little or no help addressed in our previous studies, namely the authenticity of the test setting and the integrative use of drawings. Some of the qualitative findings indicate that the reason why these two new forms of help did not yield the expected positive effect was that they conflicted with students' implicit norms, expectations, and beliefs about doing mathematics (Davis, 1989; Yackel & Cobb, 1996), especially about their appreciation of formal and informal strategies and of drawings as a valuable modelling tool. From that point of view, interventions (like the ones that were applied in the present study) that are only partial and instantaneous, and that are unable to influence or alter these more fundamental attitudes and beliefs, have little chance of success. Most likely, only a long-term classroom intervention, not only acting upon students' conceptual understanding of proportionality in a modelling context, but also taking into account the social, cultural, and emotional contexts for learning, can produce a positive effect in defeating students' overuse of linearity (see also the teaching experiment reported in chapter 5).

Second, at a more general level, the results of the present study support the warnings of several authors from the mathematics education community (e.g., Freudenthal, 1978; Treffers, 1987) to be very cautious when applying general principles and findings from cognitive and educational psychology to teaching and learning in a particular subject-matter domain like mathematics – with little regard for the specific nature of (the subdomain of) mathematics and mathematics education that is addressed in these studies (Freudenthal, 1991). Research conceived and supervised by cognitive and instructional psychologists together with experts in the domain (e.g., mathematics and mathematics education) will contribute to unravel when and how certain kinds of support that are widely accepted as effective and valuable within the community of cognitive and instructional psychologists are indeed helpful to prevent and/or overcome certain learning difficulties in a particular curricular domain like mathematics.

5. FOLLOW-UP STUDY 4: FROM INDIRECT TO DIRECT MEASURES

5.1 Rationale

In all studies reported up to now, the problems were typically constructed in such a way that it was not explicitly stated that the problems were dealing with the perimeter, area, or volume of the involved figures. Instead, they dealt with indirect measures for these magnitudes. For example, in the 'farmer Carl' problem (Table 2-1), the problem statement mentions *the time needed to fertilise* a certain piece of land, without an explicit reference to *the area* of the piece of land itself. Of course, it can be reasonably supposed that the time needed for fertilising the piece of land is directly proportional to its area, so that this time can be considered as an appropriate indirect measure for the area. Other indirect measures for the area of a piece of land could be the amount of fertiliser needed for fertilising the piece of land or the number of corn plants that can be planted on it. Analogously, the time needed to dig a ditch around a piece of land can be considered as an indirect measure for its perimeter (as would be, for example, the material cost to make a fence around the pasture), and the weight of an object as an indirect measure for its volume.

There is a good reason why we decided to use indirect measures for perimeter, area, and volume in our previous research. Let us first consider the reasoning needed to solve problems with indirect measures. Afterwards, we can compare this with the reasoning needed to solve problems with direct measures.

The following problems (about an irregular and a regular figure respectively) involve an *indirect area measure*:

> The logo of our school is an owl. If I draw the logo with a height of 50 cm, I need 110 ml of paint. When I would draw an enlarged version of the logo, having a height of 150 cm, how much paint would I need?

> Farmer Carl needs approximately 8 hours to fertilise a square piece of land with a side of 200 m. How many hours will he approximately need to fertilise a square piece of land with a side of 600 m?

In both cases, the value of the indirect area measure in the small figure is essential to find the answer (namely, *110 millilitres* of paint to draw the small logo, and *8 hours* time to fertilise the small piece of land). The only way to solve the problem about the irregular figure is by knowing that the area is multiplied by 9 if the figure is made three times as large, or by deducing/discovering this from a sketch or drawing. The problem about the square (or about regular figures in general) allows an additional method, namely by reasoning that "a square with a side of 200 m has an area of 40 000 m^2, and a square with a side of 600 m has an area of 360 000 m^2. This is 9 times more, so the time to fertilise the land is 9 times more as well". In sum, any solution procedure for problems with indirect measures necessarily focuses on the question *what happens to the* (indirect measure related to the) *area, as a consequence of the enlargement.*

Compare this with the problems rewritten with *direct measures for area:*

The logo of our school is an owl. If I draw the logo with a height of 50 cm, it has an area of 1 100 cm^2. If I were to draw an enlarged version of the logo, having a height of 150 cm, what would its area be?

Farmer Carl needs to fertilise a square piece of land with a side of 200 m. It has an area of 40 000 m^2. What would be the area of a square piece of land with a side of 600 m?

To solve the ‘owl-problem’ (about an irregular figure) with direct measures, one essentially needs to make the same reasoning as for solving the ‘owl-problem’ with the indirect measures, namely applying the knowledge that the area is multiplied by 9 if the figure is made three times as large, or by deducing/discovering this from a sketch or drawing. But by reformulating the ‘farmer Carl’ problem (about a regular figure) using direct area measures, the problem statement suddenly contains more information than needed: The last sentence of the problem is enough to solve the problem, since one can easily and immediately find the answer to the problem by just calculating the area of a square with a side of 600 m (600 m × 600 m = 360 000 m^2). So, the student does not necessarily have to reason about the impact of the enlargement on the area. This strategy is possible for squares, circles, and other regular figures. But when the problem is about an irregular figure, the only way to solve the problem is by reasoning about the increase of a perimeter/area/volume as a consequence of the enlargement (either by knowing how the perimeter/area/volume increase with the length if a figure is enlarged, or by deducing it from a sketch or drawing). Therefore, when problems are expressed with direct measures, the comparison between regular and irregular figures becomes problematic. When indirect measures are used, however, one necessarily needs to focus on the degree of increase of the area as a consequence of the enlargement of the figure, no matter what type of figure is involved in the problem.

Although, as explained just above, it was assumed that the use of indirect measures was the most valid way to observe students’ overuse of linearity in the previous studies, one could throw doubts upon the neutral character of the type of measures (direct vs. indirect) in proportional and non-proportional geometry problems, and could argue that the use of indirect measures in our previous studies might have strengthened students’ tendency to apply linear solution strategies. If that were the case, it would jeopardize to some extent the validity of our (alarming) results and our conclusions about the strength of the linearity illusion among traditionally taught secondary school students. Both theoretical and empirical indications may sustain the non-neutral character of measure types.

We first discuss a theoretical indication in favour of this argument: A rational task analysis shows that in solving problems with indirect measures (compared to solving problems with direct measures), two extra thinking steps are required to find the answer. A first additional thinking step consists of noticing whether the indirect magnitude under consideration is related to the perimeter, the area or the volume of the geometrical figure under consideration (whereas in problems with direct measures, this is explicitly stated). Second, one must realise that the relation between the indirect magnitude under consideration (e.g. the time needed to fertilise a piece of land) is directly proportional to the direct magnitude (the area of the piece of land). Because of these two extra thinking steps, more errors can occur when students solve problems with indirect measures.

Besides this theoretical indication, there are two empirical indications to be found in the research literature on mathematical problem solving, which also favour the hypothesis that indirect measures in the problem statement may affect students' answers.

First, there is the finding (coming from the general research literature on mathematical problem solving) that students sometimes tend to use 'key word strategies' (see, e.g., Verschaffel et al., 2000). In those cases, students are immediately, and mainly, led by superficial characteristics of the problem text (such as the presence of certain words) in opting for a particular mathematical operation or solution procedure. Depending on the relation between the key word in the problem statement and the solution to the problem, such 'key word strategies' can lead to a correct or an incorrect answer. In the case of non-linear geometry problems, the presence of the words 'perimeter', 'area' or 'volume' or expressions with direct area or volume units (such as cm^2 or cm^3) might remind the student to apply another strategy than the most straightforward proportional solution scheme, while the student might not be reminded to do so if an indirect measure (though proportionally related to the area or volume) is mentioned in the problem statement.

A second, more specific, indication for the role of the type of measure in the problem statement comes from the research by Rogalski (1982) on elementary school children's reasoning about lengths and areas. She reports about some students who tended to over-generalise the properties of 'unidimensional' lengths in a figure (such as the length of the diameter of a circle in metres) to 'unidimensional' measures for the area (such as the amount of paint needed to cover that circle, expressed in litres). But she observed also students attributing the same properties to direct measures for area (e.g., square centimetres - cm^2) as to direct measures for lengths (e.g., centimetres – cm), resulting in assuming a proportional relation between the area of a figure and its sides. So, although the studies by Rogalski (1982) do not point in one specific direction with respect to the type of measure in the problem statement, her observations indicate at least that the type of measure can possibly influence students' thinking in such a problem situation.

All these indications indeed point in the direction that the use of problems involving indirect measures in our previous research might have strengthened their tendency towards unwarranted proportional reasoning and therefore might have had a significant negative impact on students' performances. In order to test this hypothesis, and in order to validate the findings and conclusions of our earlier investigations on students' overuse of linearity, a new empirical study was conducted in which the type of measure in the problem statement was experimentally manipulated. This fourth follow-up study is reported next.

5.2 Method

Follow-up study 4 aimed at investigating the influence of the nature of the measures in proportional and non-proportional geometry problems. For this purpose, a paper-and pencil test was administered to 145 secondary school students aged 15-16, coming from two different schools for general secondary education in Flanders.

All participants received a test with two problems: one proportional problem about the perimeter of an enlarged irregular figure (where the perimeter of the smaller version was given and the larger was unknown) and a non-proportional problem about the area of that enlarged irregular figure. These problems were presented in random order. Besides giving the answer, students were asked to write down their calculations, drawings, and anything that they did to solve the problem. We opted for using irregular figures for reasons explained above. Only for irregular figures, a problem can be formulated with direct measures, while maintaining the need to reason about the effect of a similar enlargement on perimeter or the area. A regular figure – such as a square or a circle – would lead to redundant information in the word problem if direct measures were used, so that students would not necessarily need to calculate the effect of the enlargement on the perimeter or area.

Table 3-10. Examples of word problems used in the conditions with direct measures (D-condition) and indirect measures (I-condition)

	D-condition	I-condition
Introduction	In January 2002, the Euro was introduced. Several Belgian schools worked around this event. In some of those schools, students made a chalk drawing of the map of Belgium on the playground.	
Perimeter problem (proportional)	The students in school A made a small drawing: the Belgian map had	The students in school A made a small drawing: the map had a

	D-condition	I-condition
	a height of 2 m. In the math lesson, the students carefully determined the perimeter of that map. It would be 11 m. Students in school B made a much larger version of the map. It was 6 m high. What would be the perimeter of this map?	height of 2 m. Afterwards, they put 20 eurocent coins on the chalk lines of the map. They needed 3 kg of coins to do that. Students in school B made a much larger version of the map. It was 6 m high. How many kg of coins would they need to put on the chalk lines of their map?
Area problem (non-proportional)	In both schools, they also tried to determine the area of the map. Students in school A calculated that the area of their map was 3 m^2. What would be the area of the map in school B?	In both schools, the students also paved the whole map of Belgium with 10 eurocent coins, turning it into a coin carpet. In school A, pupils needed 40 kg of coins to do that. How many coins would the students in school B need?

Two mathematically equivalent versions of the test were developed, and students were randomly assigned to one of two conditions. Half of the students received a test with problems explicitly mentioning the terms perimeter and area and using the direct measures for these magnitudes (D-condition). The other students received the test with similar problems containing only the indirect measures for the perimeter and the area of the involved figure (I-condition). Table 3-10 gives examples of the 'direct' and 'indirect' items, together with the introduction to the problems, as they were presented to the students. To guarantee that the participants would make a correct interpretation of the situations described in the word problems (i.e. an enlargement in which the figure keeps the same shape), the test also contained an image of the small and the enlarged irregular figure (see Figure 3-3).

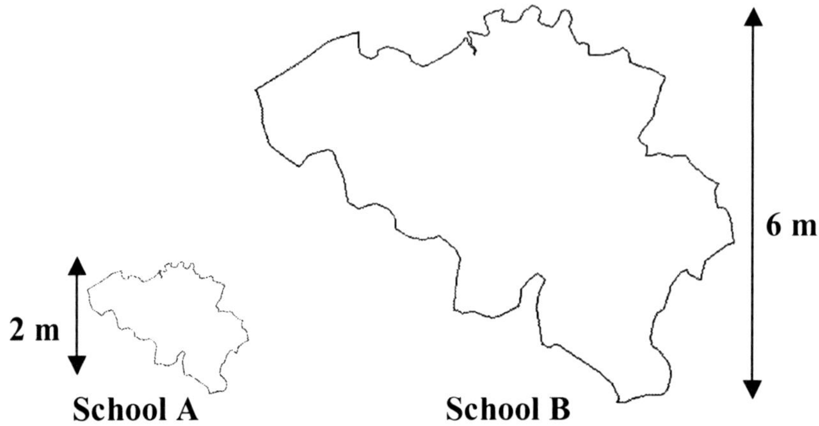

Figure 3-3. Image accompanying the word problems in both conditions

Students' solutions to both problems were analysed in two ways. First, it was determined whether the answer to the problem was correct or incorrect. These performance scores were analysed by means of a '2 × 2' repeated measures ANOVA, with 'Proportionality' (proportional vs. non-proportional) and 'Condition' (D-condition vs. I-condition) as the independent variables and the number of 'Correct answers' as the dependent variable.

A second, qualitative analysis consisted of determining the solution strategy that was applied to solve the problem. For this purpose, a qualitative categorisation scheme was developed, based on the literature on proportional reasoning and on previous research findings. It was further refined during a first round of analysis on a sample of the protocols. The inter-rater agreement for this categorisation (based on a sample of 40% of the protocols by two of the researchers) was kappa = 0.802, which is considered as sufficient for this type of analysis (Landis & Koch, 1977). Using this categorisation scheme, all correct and incorrect solution protocols were assigned to one of the following four categories. These categories are illustrated for the direct non-proportional problem mentioned in Table 3-10.

A first category refers to solutions that rely on proportionality to find the solution. Of course, such proportional solution strategies are correct for the proportional (perimeter) problem, but not for the non-proportional (area) problem. Three major subcategories are distinguished among the proportional solution strategies:

- *Internal ratio*, i.e. reasoning that the ratios of the heights of the two maps (2 m / 6 m = 3) should apply to the areas of the maps too (so 3 m^2 × 3 = 9 m^2).

- *External ratio*, i.e. reasoning that the ratio between the height and the area in the first map (2 m / 3 m^2) should also hold for the second map.
- *Rule of three*, i.e. reducing one of the quantities in the problem to its unit, and then deducing the unknown value (applied on the exemplary problem: 3 m^2 for a map of 2 m high, so 1.5 m^2 for a map per meter of the height; this means that a map of 6 m high has an area of $6 \times 1.5\ m^2 = 9\ m^2$.

A second category consists of solutions in which the 'general principle' governing the similar enlargement of geometrical figures is applied: If a figure is enlarged *r* times, the perimeter is enlarged *r* times as well and the area is enlarged r^2 times.

Third, there are solutions named 'reducing the figure'. In these solutions, the irregular figure under consideration (e.g., the outline of a map of Belgium) is reduced to a more regular figure such as a rectangle or a right-angled triangle. Such a solution then either proceeds by working on the drawing (for example 'tiling' the obtained large right-angled triangle with smaller versions of that triangle) or by reasoning with the formula for calculating the area of the regular figure (e.g. for a rectangle: small area = height × width, so large area = (3 × height) × (3× width) = (9 × small area)), or a combination of both.

The fourth category contains all remaining solutions such as unanswered problems and (sometimes correct, but mostly incorrect) solution processes that were difficult to understand or to categorise in one of the other three categories.

It is important to remark here that correct and incorrect solutions were categorised by means of this scheme, but correctness of an answer is not independent of the solution category. Solutions by means of proportionality were in principle correct for proportional problems, but, of course, always incorrect for the non-proportional problems. Solutions in which the irregular figure is reduced to a regular one could be either correct or incorrect for both types of problems, and solutions relying on the 'general principle', were generally correct for both types of problems.

5.3 Research questions and hypotheses

The key question in this study was whether the data support the general hypothesis that students' performances on and solution procedures for linear and non-linear geometry problems are different, depending on whether these word problems are formulated in terms of direct or indirect measures for the perimeter or area.

Based on that question and on the above-mentioned theoretical considerations, one would predict that the students would perform better on problems in which the measure of perimeter/area/volume is explicitly mentioned (i.e. better performances in the D-condition than in the I-condition), and especially that they would less often apply a proportional solution method when this is not applicable for the problem (i.e. better performances on the non-proportional problems, due to less proportional solution strategies).

Although all solution strategies from the qualitative analysis scheme (see the 'Method' section) are, in principle, applicable in both the D- and the I-condition, the hypothesis also allows to make specific predictions about the different strategies that would be actually applied in both conditions. Due to the presence of key words (perimeter or area) and direct perimeter or area measures, one would expect that the D-condition would yield more frequently 'content-specific' problem-solving strategies related to (determining the) perimeter and area than the I-condition (where these key words and direct measures are absent). Because the problems in the test deal with irregular geometrical figures, such 'content-specific' strategies could either consist of 'applying the general principle' of the effect of an enlargement on perimeter and area (the second category in our scheme), or of 'reducing the irregular figure' to a regular one (such as a rectangle or a right-angled triangle) and calculating the area, applying formulas or using a sketch or drawing (i.e. the third category in our scheme). Because in the I-condition, the key words (perimeter or area) and their corresponding measures are absent, we expected more solution strategies relying on proportionality (internal and external ratio, and 'rule of three') in that condition than in the D-condition: These proportional strategies have a status of more or less 'generally applicable' strategies, in contrast with the earlier-mentioned 'content-specific' strategies.

5.4 Results

Table 3-11 presents an overview of the performance of the students on the proportional and the non-proportional word problems in general and for the two research conditions separately.

First, students again performed much better on the proportional item than on the non-proportional item. For the two conditions together, about 85% of the students answered the proportional item correctly, whereas only 23% gave a correct answer to the non-proportional item. This result was in line with our previous research findings confirming students' overwhelming tendency to improperly apply proportional solutions on non-proportional word problems.

Second, we found no differences in performance between the two research conditions, either for the proportional item (involving the perimeter), or for the non-proportional item (involving the area). Therefore, the hypothesis that students would perform better on non-proportional problems if they were expressed with direct measures for the area was not confirmed.

Table 3-11. Mean performances (and standard deviations) of the students on the proportional and non-proportional problem in the I-condition and in the D-condition

	Direct		Indirect		Total	
	Mean	SD	Mean	SD	Mean	SD
Proportional	0.792	0.410	0.876	0.331	0.848	0.360
Non-proportional	0.208	0.410	0.237	0.427	0.228	0.421
Total	0.500	0.503	0.557	0.498	0.535	0.499

Despite the absence of an effect of 'Condition' (or of the interaction with 'Proportionality'), the question remains whether students in the D-condition applied other solution strategies than the students in the I-condition, so that the solution process could have been affected by the type of measure while the performances were the same. To answer this question the above-mentioned qualitative analysis was conducted. Table 3-12 gives an overview of the solution strategies that were applied by the students in each condition to solve the proportional and non-proportional word problem.

Table 3-12. Overview of the solution strategies (in %) used by the students to solve the proportional and non-proportional problem in the D- and I-condition

Type of solution strategy	Direct			Indirect		
	Proportional	Non-proportional	Total	Proportional	Non-proportional	Total
Proportionality	54.2	66.7	60.4	80.4	64.9	72.7
Internal ratio	45.8	60.4	53.1	66.0	51.5	58.8
External ratio	6.3	6.3	6.3	5.2	5.2	5.2
Rule of three	2.1	0.0	1.0	9.3	8.2	8.8
Reducing figure	22.9	16.7	19.8	4.1	12.4	8.2
General principle	0.0	4.2	2.1	0.0	12.4	6.2
Other	22.9	12.5	17.7	16.5	10.3	12.9
Total	100.0	100.0	100.0	100.0	100.0	100.0

This table shows that most of the problems in both conditions were solved by a solution strategy relying on proportionality, mostly an 'internal ratio' strategy (which is also the most widely used proportional solution strategy among students in general, see e.g., Tourniaire & Pulos, 1985). This explains why most students performed well on the proportional item (they correctly reasoned that the perimeter was tripled since the height was tripled), but failed on the non-proportional item (since the area and the number of coins to cover the figure were not tripled). Only a minority of the students thought of applying an approach whereby the irregular figure under consideration was reduced to a regular one, and even fewer students applied the general principle.

A comparison of the D- and the I-conditions showed that there were some small but interesting differences with respect to the strategies used by the students. As expected, proportional strategies such as the 'internal ratio' or the 'rule of three' were more often applied in the I-condition (73%) than in the D-condition (60%). When students in the I-condition recognized the non-proportional character of the area problem, this happened sometimes because they knew and activated the general principle (12%), sometimes because they reduced the irregular figure to a regular one (12%). In the D-condition, however, non-proportional strategies for the area problem more often consisted of applying a 'reducing the figure' strategy (17%) (which was in line with our expectations), whereas the 'general principle' (4%) was a less often applied non-linear strategy (which was contrary to our prediction). The presence of direct measures for perimeter and area seems to trigger other strategies in some students. It reminds them of applying previously acquired knowledge about areas of rectangles or triangles, of working on a provided drawing etc., whereas problems with indirect measures elicit more often more 'holistic' or more 'general' approaches for solving word problems (in this case, the application of proportionality).

5.5 Conclusions and discussion

In our earlier studies, it was repeatedly found that many students believe that if a figure is enlarged by a factor r, not only the perimeter but also the area and volume of that figure increase with that factor r. In the problems used in these studies, however, the magnitude under consideration typically was an indirect – though proportionally related – measure for the perimeter, area, or volume, e.g. the weight of an object as an indirect unit for measuring its volume. We assumed that this would have no significant influence on students' solutions. The fourth follow-up study explicitly addressed this assumption by testing whether the use of indirect measures in non-linear geometry problems (as done in our earlier studies) strengthens students' illusion of linearity, and thus may have influenced the research findings on this misconception.

In this study, the measures in the problem statement were experimentally manipulated. Half of the students solved two items involving direct measures for perimeter and area, while the others solved isomorphic items with indirect measures. A comparison was made of students' performances as well as of their solution strategies. We found no significant differences in the performances on the two types of problems. Apparently, the type of measure used in the problem statement has no significant influence on students' performance in general, and on the occurrence of improper proportional reasoning in particular. This seems to confirm the validity of the findings of the earlier studies on students' overuse of linearity. A qualitative analysis of the underlying solution strategies, however, provided some interesting differences. For items with indirect measures, more students applied a strategy based on thc general application of linearity, whereas items with direct units for perimeter and area elicited more content-specific strategies such as working on the graphical representation and the application of formulas for perimeter and area.

We conclude with an important remark. The aim of this last follow-up study was to investigate whether a test involving *indirect* measures provides equally valid data to study the illusion of linearity as a test involving *direct* measures. Since both tests yielded comparable data, they both seem valid for the purposes of our research. But this does not imply that there is no difference in the difficulty of solving problems with direct and indirect measures for perimeter, area, and volume whatsoever. The low percentage of students applying the 'general principle' (cf. Table 3-12) indicates that, apparently, only a very small minority of them (and of students in the previous studies) had really understood the effect of a linear enlargement on area and volume. For these students, the presence of direct or indirect measures in the problem statement made no difference. But for students who possessed this insight, the measures in the problem statement might have had the hypothesized influence.

6. GENERAL SUMMARY AND DISCUSSION

The two studies reported in chapter 2 convincingly demonstrated the occurrence and strength of students' overuse of linearity with respect to problems about the relations between the linear measurements and the area of similar plane figures. However, it remained unclear to what extent the results of those studies were affected by aspects of the experimental setting and of the way in which the problems were presented and formulated, namely the collective administration of paper-and-pencil tests consisting of proportional and non-proportional missing-value problems under different experimental conditions to large groups of 12–13- and 15–16-year old students. Therefore a series of four follow-up studies was conducted with the same age groups and following roughly the same methodology. Each of these studies tested a particular hypothetical explanation for students' tendency to give linear answers. The greater the improvement in students' test scores (on the non-proportional items) in these follow-up studies, the more evidence we would have that the alarming results obtained in the studies reported in chapter 2 were not as frightening as initially thought.

In Follow-up study 1, the effect of two different scaffolds was investigated: (1) a metacognitive scaffold in the form of an introductory warning before the start of the actual test (to avoid students approaching the test with the expectation that it only contained routine tasks) and (2) a visual scaffold in the form of ready-made drawings on squared paper (as such kind of drawings provides more suitable reference points for determining lengths and areas in an informal way). Both scaffolds yielded significant but small effects on students' performance. The percentage of correct answers on the non-proportional items rose from 12% to 18% as a consequence of the metacognitive scaffold, and from 13% to 17% as a consequence of the visual scaffold. Offsetting the better results on the non-proportional items, the results on the proportional items slightly decreased as some students started to apply non-proportional methods to these items too.

In Follow-up study 2, the role of the traditional missing-value formulation of the word problems was investigated as students in the previous studies might have linked this format quasi-automatically to the application of a proportionality scheme. This was realised by confronting students with mathematically equivalent problems formulated as comparison problems (e.g., for the non-proportional item mentioned above: "Farmer Carl fertilised a square piece of land. Tomorrow, he has to fertilise a square piece of land with a side *being three times as big. How much more time* will he approximately need to do this?"). It appeared that students who received comparison problems performed significantly better than students who received missing-value problems (41% and 23% respectively). Nevertheless, more than half of the rephrased non-proportional problems were still solved erroneously. Moreover, a better performance in the comparison task group again went together with a worse score on the proportional items.

The third follow-up study focused on two new experimental manipulations, namely (1) making the problem context more authentic (as students' weak performances in the previous studies might have been caused by the unattractiveness or inauthenticity of the problem context) and (2) making the drawing activity truly part of the problem-solving process (as students in the previous studies might have seen the instruction to make a drawing or the provided drawings as unrelated to the subsequent task). The first scaffold was realized by showing students before the test some video fragments telling the story of Gulliver's visit to the isle of Lilliputians – where all lengths are 12 times smaller – and linking all items to these fragments. The second scaffold was realized by asking students to draw a reduced copy of the geometrical figure in the problem before they got the numerical data to solve it. Contrary to the expectations, both scaffolds yielded a significant effect on students' performances on the non-proportional problems (a decrease from 41% to 25% correct answers for the manipulation of the authenticity and a decrease from 44% to 23% for the integrated drawing activity effect). Several explanations for these unexpected results were discussed. Generally spoken, the absence of positive effects for the two scaffolds was explained by the fact that these scaffolds conflicted with students' implicit norms, expectations and beliefs about doing mathematics in general and about applying formal and informal strategies (and using drawings) as a modelling tool.

Follow-up study 4 empirically tested the role of the measures (direct vs. indirect) that are used in problem formulations (a role that was assumed to be neutral in previous studies). Therefore, the measures in the problem statement were experimentally manipulated. It was confirmed that the type of measure did not significantly affect students' performances, but some interesting differences with respect solution strategies were revealed: For problems with indirect measures, more students applied a strategy based on the general application of linearity, whereas problems with direct units elicited more content-specific strategies such as working on the graphical representation and the application of formulas for perimeter and area.

To sum up, all studies reported so far showed that the vast majority of 12–16-year-old students failed on non-proportional word problems about length, area, and volume of similar plane figures, because of their alarmingly strong tendency to apply proportional reasoning 'anywhere'. Although various experimental manipulations yielded significant positive effects on students' performance on non-proportional items, these effects were disappointingly small. Moreover, each time the performances on the non-proportional items improved due to one of the above-mentioned forms of help, this was paralleled with worse performances on the 'easy' proportional items. Most often, errors on these proportional items were due to the unwarranted application of non-proportional solution strategies to these items too, suggesting that students' emerging non-proportional reasoning skills were still fragile and unstable.

Chapter 4

AN IN-DEPTH INVESTIGATION

1. INTRODUCTION

Whereas all our previous studies involved quantitative and qualitative analyses of large groups of students' reactions to paper-and-pencil tests, the fourth chapter reports an in-depth investigation by means of individual semi-standardised interviews performed with small numbers of 12–13- and 15–16-year old students (see also De Bock, Van Dooren, Janssens, & Verschaffel, 2002a). In fact, the administration of collective tests to large groups of students under different experimental conditions did not yield adequate information on the *problem-solving processes* underlying improper proportional responses. Moreover, it remained largely unclear what aspects of students' knowledge base were responsible for the occurrence and strength of this phenomenon and how these aspects relate to other more general misconceptions and buggy rules identified in the literature. Therefore, we made this shift in our methodology by doing in-depth interviews with individual students who fall into the 'proportionality trap'. The interview study that will be reported here was preceded by a pilot study aimed at the development, try-out, and definitive design of different aspects of this interview technique (see De Bock, Van Dooren, Verschaffel, & Janssens, 2001).

2. METHOD

Twenty 12–13-year olds and 20 15–16-year olds participated in the study. The participants came from a school located in a medium-sized Flemish town and were equally divided over most of the study streams of general secondary education. The number of boys and girls in the school as well as in our sample was more or less the same. All interviews were audio-taped and the interviewees had paper, writing materials, a ruler and a scientific calculator at their disposal. The interviews consisted of five phases: After a short introduction, the interviewee was confronted with a non-proportional problem (Phase 1), followed by four subsequent forms of help to solve that problem (Phases 2, 3, 4, and 5). The interview stopped in the phase where the students discovered the non-linear nature of the problem and gave the correct response. If no form of help proved to be successful, the interview stopped at the end of Phase 5. The forms of help provided in Phases 2 to 5 aimed at eliciting a cognitive conflict in students who fell into the 'linearity trap'. As explained earlier in chapter 3, eliciting a cognitive conflict is a well-known method to create cognitive disequilibrium in a learner and to lead him or her to the discovery and development of new ideas (Forman & Cazden, 1985; Limón, 2001; Perret-Clermont, 1980). In the present study, the cognitive conflict was evoked by presenting parts of the problem-solving protocol of a fictitious peer who proposes a (correct) non-linear solution to the problem. From Phase 2 to 5 this cognitive conflict in the interviewee was gradually increased by providing more and stronger evidence for the non-linear solution. Just to be perfectly clear, the main purpose of the cycle of cognitive conflicts used in the interviews was to unravel students' thinking processes by ascertaining how they reacted to particular kinds of help, and was thus not meant as a didactical trajectory. Preventing or remedying students from falling into the 'linearity trap' would require another approach, such as the instructional units 'Gulliver' developed by Treffers (1987) or 'With the giant's regards' developed by Streefland (1984) or the lesson series that we developed and that will be described in chapter 5. But the development and/or evaluation of such an instructional unit was not the focus of the present study.

Technically speaking, the interviews can be characterised as semi-standardised: The interviewer-researcher followed a pre-determined scheme for the global development of the interview, asked specific standardised questions, but left enough room for spontaneous reactions of the interviewee and tried to respond to these reactions in a flexible way (Ginsburg, Kossan, Schwartz, & Swanson, 1982). A potential weakness of this methodology is the social effect that may bias the results. As the interviewer is consistently trying to raise more and more doubts in the student's mind, a student who has less understanding might be inclined to change his or her mind purely on the basis of perceived social pressure and not because of an (increased) understanding of the problem. Whereas this factor may lead to an artificial increase in the number of correct reactions, 'self-defence' reactions may result in the opposite effect. Due to such a tendency to demonstrate firmness or the (mis)belief that in problem solving it is always better to stick to one's initial response (Verschaffel, De Corte, & Vierstraete, 1997), some students may stick to their original answer, even though they realise that this answer is untenable (see also De Bock et al., 2001). We were on the alert for this aspect of the 'experimental contract' (Greer, 1997) by the above-mentioned studies and, therefore, the interviewer stressed that, in every phase of the interview, the student was free to revise his answer whenever he felt this was necessary, since the interviewer was only interested in whether the student could give the correct answer *by the end* of the interview and in how this final answer was obtained. So, we made every possible effort to create an experimental 'climate' in which a student would not resist changing his solution as soon as he thought or realised it was incorrect (and, of course, resist changing it when still convinced about its correctness).

We now describe the five interview phases in more detail. A summary of each phase is given in Figure 4-1. The interview started with a standardised *Introduction* explaining to the student that the interview was part of a research project on mathematical problem solving. The student was told that he would be asked to solve one single problem. The interviewer stressed that, during the interview, he was always free to revise his solution whenever he felt this was necessary.

In *Phase 1* of the interview, the student received one non-proportional word problem about the enlargement of an irregular two-dimensional figure. To guarantee an appropriate and uniform interpretation, the problem was accompanied by a drawing of the original and enlarged figure. Previous research had shown that the vast majority of 12–16-year old students solve problems about the area of enlarged irregular figures in a proportional way, even when these problems are accompanied by ready-made drawings (see chapter 2). To prevent students getting inside information about the problem and/or its correct answer from their classmates being interviewed earlier, we used four isomorphic versions of the same problem, one of which is given in Figure 4-1. The student was first asked to read the problem aloud and to 'think aloud' (Ginsburg et al., 1982) while solving it. At the moment when the student finished his first reading of the problem, a chronometer was started to measure the response time. When the thinking-aloud protocol did not yield sufficient information about the student's thinking process, the student was asked to explain how his answer was found. Then, the student was asked to indicate how sure he was about the correctness of that answer, by choosing a position on a five-point scale (from 'certainly wrong' to 'certainly correct'). When a student did not indicate 'certainly correct', the interviewer asked why he was not absolutely sure, if there was anything that did arouse doubts, and if he had considered alternatives. At the end of Phase 1, the student was asked to explain *why* he thought the problem had to be solved in that way. When a student could not justify his answer, the interviewer made use of a 'teaser' consisting of a nonsensical additive solution for the problem and asking the student why his solution was better than this one (referring to the problem given in Figure 4-1, the 'teaser' was: "The second Father Christmas is 168 cm – 56 cm = 112 cm higher, thus Bart will need 112 ml more paint").

In *Phase 2*, we tried to raise a first, weak form of cognitive conflict in the students who had solved the problem incorrectly by means of proportional reasoning. This was done by confronting them with a fictitious frequency table presenting an overview of the answers given by a group of peers (see also Figure 4-1). The fictitious table contained two major answer categories. For the Father Christmas problem in Figure 4-1, for instance, the table indicated that 43% of the peers answered 18 ml (which is the incorrect, linear answer given by the student himself), but another 43% answered 54 ml (which is the correct, non-linear answer). The remaining 14% in this fictitious frequency table gave other answers or could not give any answer at all. At first, no questions were asked and we observed to see if the student spontaneously searched for the origin of the equally popular alternative. If the frequency table did not elicit a spontaneous reaction, the interviewer asked if the student had any idea where the alternative non-linear answer did come from and if this alternative did raise some 'seeds of doubt' about his initial answer. Finally, the student was asked which answer he preferred: the initial answer or the alternative that emerged in the peer group. After the student made his decision, the interviewer once more asked for a justification.

For the students who did not abandon their initial linear answer at the end of the second phase, a stronger conflict was elicited in *Phase 3*. In this phase, we gave the argumentation of a fictitious peer from the 43% who answered the problem correctly. For the example listed in Figure 4-1, the following argumentation was given: "One student told me that if the Father Christmas becomes three times as high while keeping the same shape, not only is his height multiplied by 3, but also the *width* has to be multiplied by 3, so that you have to multiply the amount of paint by 9". Moreover, the calculation '9 × 6 ml' was written down next to the answer '54 ml' in the frequency table. If the student did not react spontaneously, the interviewer asked if the argumentation of the peer did (not) raise doubts about his initial answer. Finally, the student was invited once again to indicate his preference between the linear and the non-linear answer. Students who did not exchange their original linear answer for the correct non-linear one, went to Phase 4.

In *Phase 4* an even stronger cognitive conflict was installed by demonstrating (visually) the reasoning behind the argumentation of the fictitious peer. Therefore, the interviewer showed again the two Fathers Christmas of Figure 4-1, but now inscribed in exactly fitting rectangles (see Figure 4-1). He explained that the peer who multiplied by 9 drew the rectangles around the Fathers Christmas and then saw that it enlarges 3 times in both dimensions. So, the amount of paint needed for the big Father Christmas is $3 \times 3 = 9$ times the amount needed for the small one. This intervention was inspired by the study reported in chapter 2 in which students more easily discovered the two-dimensional impact of an enlargement in regular than in irregular figures. Once more, after leaving room for spontaneous reactions, the interviewer asked if the solution strategy of the peer did (not) raise doubts about his initial answer. Finally, the interviewee was invited to indicate his preferential answer. If this preferential answer was maintained as the incorrect linear one, the student arrived in the last interview phase.

The strongest cognitive conflict was evoked in *Phase 5* wherein we created an explicit link with the area of regular and irregular figures. The student was consecutively asked (1) to calculate the area of the rectangles wherein the two Fathers Christmas were inscribed, (2) to compare the area of these two rectangles, (3) to compare the area of the two inscribed Fathers Christmas, and (4) to compare the amounts of paint needed for painting them (for the exact phrasings, see Figure 4-1). If this fourth scaffold proved to be ineffective, the interviewer asked the student to compare the amounts of paint needed for painting the two rectangles. Because every step in Phase 5 could be helpful, the interviewer left room for spontaneous reactions after each step. At the end, the interviewee stated his definite preference and whatever this answer was, the interview stopped here.

Phase 1 – Solving a word problem

Bart is a publicity painter. In the last few days, he had to paint Christmas decorations on several store windows. Yesterday, he made a drawing of a 56 cm high Father Christmas on the door of a bakery. He needed 6 ml paint. Now he is asked to make an enlarged version of the same drawing on a supermarket window. This copy should be 168 cm high. How much paint will Bart approximately need to do this?

Bakery door **Supermarket window**

Phase 2 – Weak form of cognitive conflict

"Last week, we gave this problem to pupils in another school. This table shows how they answered it."

Answer	*Number of pupils*
18 ml	43 %
54 ml	43 %
Other	14 %

Phase 3 – Argumentation of fictitious peer

"One student told me that if the Father Christmas becomes three times as high while keeping the same shape, not only is its height multiplied by 3, but also the width has to be multiplied by 3, so that you have to multiply the amount of paint by 9.

$$9 \times 6 \text{ ml} = 54 \text{ ml"}$$

Phase 4 – Solution process of fictitious peer

"The student answering 54 ml also explained me how he found that answer. He drew rectangles around the Fathers Christmas, and then he saw it enlarges 3 times in both dimensions".

Bakery door

Supermarket window

Phase 5 – Explicit link with area measurement

- "Can you calculate the area of the two rectangles?"
- "How much larger is the area of the large rectangle compared to the small one?"
- "How much larger is the area of the large Father Christmas compared to the small one?"
- "How much more paint do you need to paint the large Father Christmas?"

Figure 4-1. Summary of the interventions in each phase of the interview

3. RESULTS

To illustrate how the interviews ran and to give an idea of some of the students' problem-solving processes and the observed reactions, we will first present the interviews of Tommy (aged 12) and Anne (aged 16). We selected these two students because their answers during the interviews were most representative of the reactions of the whole group of students. After that, we provide a systematic overview of the reactions of the different students in each phase of the interview.

3.1 Two sample interviews

The interview of Tommy (12). *The researcher offered Tommy a sheet with the Father Christmas problem and the drawings presented in Figure 4-1. The interview proceeded as follows:*

Tommy: [reads the problem aloud] Ehm, wait, let me look for the numbers … I see, the height changes from 56 cm to 168 cm. That means multiplying by three. So, I have to multiply the paint by three too. [writes down the scheme in Figure 4-2]

The answer is 18 ml. Bart will need about 18 ml for the large Father Christmas.

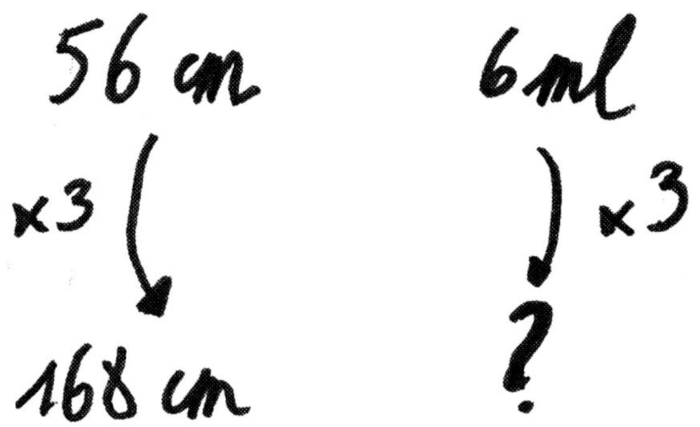

Figure 4-2. Tommy's solution scheme

Interviewer:	Can you elaborate a bit on your answer? Why did you solve the problem in that way?
Tommy:	[silence] Eh. I don't know. I just solved it that way.
Interviewer:	Why did you multiply the amount of paint by three?
Tommy:	It is so logical. It can't be done else, can it? The Father Christmas becomes *higher*, so you need *more* paint. And it becomes *three times* higher, so you need *three times more* paint. It is as simple as that!

Interviewer: Can you express how sure you are that your solution is correct?

Tommy: I'm very sure. My answer is certainly correct. It is an easy problem. I just used the three numbers in the problem and the formula.

At this moment, the interviewer provided the fictitious frequency table presenting an overview of the answers given by a group of peers (see also Figure 4-1), and continued as follows (Phase 2):

Interviewer: Take a look at this... We gave that same problem to a large group of students in another school. This table shows how they answered it. You see: 43% gave the same answer as you did, namely 18 ml. But there was another 43% of the students who answered 54 ml. Who is right?

Tommy: [immediately] No, no, that is impossible. I have calculated it and it surely is 18 ml. How do they get 54 anyway? I will try ... [subtracts 54 from 168, tries a few combinations of 54, 168, 6 and + - × and :] Oh wait, I see! They multiplied by three two times. You see, they made a mistake. My answer is correct.

Since Tommy persisted, the interviewer moved to Phase 3 of the interviewing procedure:

Interviewer: One student from that other school who answered 54 ml, explained how he solved the problem. That student argued to me that if the Father Christmas picture is enlarged three times, not only the height but also the width is multiplied by three, so that you need nine times more paint...

Tommy: Oh, that student uses the picture. I didn't use the picture. I just looked at the text of the problem. And the text only mentions the height.

Interviewer: And what if you would look at the drawing too?

Tommy: 18 ml is and remains my answer. That student makes it too complex. My answer is easier.

At this point the interviewer made the next step in the helping procedure (Phase 4). He presented the solution process of the fictitious peer (see also Figure 4-1), together with the following explanation:

Interviewer: Some students who answered 54 ml solved the problem in this way. They drew a rectangle around both figures and found that the figure enlarges three times in both directions: the height and the width. What do you think of that strategy? What solution do you prefer now?

This intervention led to the following interaction with Tommy:

Tommy: But the problem says nothing about the width at all. The problem is about the height and the amount of paint.

Interviewer: And what about the strategy with the rectangles?

Tommy: What they say about the rectangles is correct: the rectangle enlarges in two directions. But within the rectangles, there is an irregular figure, and that's quite different. There are white parts in the rectangles, and they are larger for the large Father Christmas. [points at the empty spaces within the rectangles]

Finally, the interviewer made an explicit link with area measurement (Phase 5).

Interviewer: Can you calculate the areas of the rectangles and compare them?

Tommy: [calculates areas correctly] This one is nine times larger.

Interviewer: And how about the Father Christmas drawings?

Tommy: [immediately] No, that's weird reasoning. You make it three times larger, so you need three times more paint. You make it too difficult whereas mathematics is logical. My answer is that he needs 18 ml to paint the large Father Christmas.

At this stage, no further help was given to Tommy, and the interview was stopped.

The interview of Anne (16). *Anne was given the same Father Christmas problem as Tommy. Her interview went as follows:*

Anne: [reads the problem] Oh I see. You need 6 ml for painting 56 cm. Then I can calculate how much you need for 1 cm. [divides 6 by 56 with a calculator] Here it is! You need 0.1071 ml for 1 cm. Then I multiply by 168, because the large Father Christmas is 168 cm. [calculates and writes down the calculations shown in Figure 4-3] You need 18 ml for the larger version of the Father Christmas.

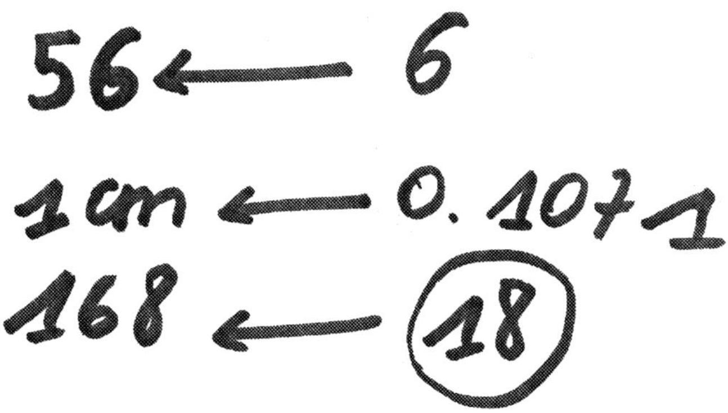

Figure 4-3. Ann's calculations

Interviewer: Can you elaborate a bit on your answer? Why did you solve the problem in that way?

Anne: It just works. I don't know why. I do such problems always like that. You just find out how much you need for 1 cm, and the rest follows automatically.

Interviewer: How sure are you that your solution is correct?

Anne: I'm not completely sure, because I haven't carefully read the problem. And I might have made a calculation error. But I think I did what was expected, and I used all three numbers in the problem.

Since Anne gave the proportional answer, the interviewer provided the first hint, namely offering the frequency table (Figure 4-1) and pointing to the equally supported alternative answer. Anne reacted in this way:

Anne: What? 54 ml? I think that is quite a lot. I think my solution is more logical. Besides, it's always better to stick to your first solution.

Interviewer: A student in that other school argued to me that if the Father Christmas picture is enlarged three times, not only the height but also the width is multiplied by three, so that you need nine times more paint. That's why he answered 54 ml…

Anne: [interrupts] No, I don't think so! Indeed, it becomes three times higher and three times wider! But that means exactly that the amount of paint is also multiplied by three. The 6 ml is for the whole Father Christmas, not only for the height. And 18 ml is for the whole large Father Christmas. You see? This area fits three times in this area [points roughly to the small and large figure], so you need three times more paint here.

Since Anne persisted in her proportional solution the next hint was provided. The interviewer showed and explained the solution strategy using rectangles (as also shown in Figure 4-1). When seeing these figures, Anne immediately decided to change her answer:

Anne: Oooh yes, now I see it! It is indeed nine times larger, because the small rectangle fits nine times in the larger one. With help of these rectangles, I suddenly see it ... The answer is 54 ml.

Interviewer: Can you explain why you answered 18 ml at the beginning? You seemed rather convinced of your answer …

Anne: Yes, but my answer seemed so logical: three times larger, three times more paint. I looked at the text and I knew immediately what I had to do. If I had taken a look at the drawings, I might have noticed that my strategy wouldn't work. But I just focused on the calculations.

3.2 Overview of students' reactions

General observations. Table 4-1 presents the number of students who chose the correct answer in each phase. As expected, the tendency to give a linear answer was strongly present in both age groups. Initially, only two students, both 15–16-year olds, solved the problem correctly. The other 38 students gave an erroneous, linear answer in Phase 1 of the interview. The subsequent cognitive conflicts of Phases 2 to 5 proved to be effective for, respectively, one, seven, five, and three 12–13-year olds and for, respectively, one, seven, four, and two 15–16-year olds. By the end of the interview, four students in each age group had not exchanged their wrong, linear answer for the correct, non-linear one. We now look at each phase in more detail, particularly from a qualitative point of view. Differences between the two age groups will only be described if relevant.

Table 4-1. Absolute and cumulative number of students who chose the correct answer in each phase

Age group	*N*	Phase 1	Phase 2	Phase 3	Phase 4	Phase 5
12–13-year	Absolute	0	1	7	5	3
($N = 20$)	Cumulative	0	1	8	13	16
15–16-year	Absolute	2	1	7	4	2
($N = 20$)	Cumulative	2	3	10	14	16
Total	Absolute	2	2	14	9	5
($N = 40$)	Cumulative	2	4	18	27	32

3.3 Phase 1: Solving the word problem

The mean response time after the first reading of the word problem was 98 seconds. Without the response times of the two 15–16-year olds who gave the correct non-linear answer in Phase 1 (125 and 130 seconds), the mean is still 97 seconds. This is rather high, considering that most students used very simple and straightforward proportional calculations. It is caused, however, by some students having difficulties with manipulating the available calculator and by some others who had to re-read the problem because they had not read it carefully enough the first time, or because they had not written down nor memorised the relevant numerical data. None of the 40 students made a drawing or any other kind of external representation involving more than writing down the three given numbers.

Except for the two 15–16-year olds who gave the correct non-linear answer in this phase, all students calculated the amount of paint needed in a proportional way. Only one of these 38 students expressed some doubts about this linear approach ("You need more data, for instance the width"), but this did not affect his solution. Strategies for finding the linear solution were, in order of importance, the internal ratio strategy (i.e. using the ratio between the heights of the Fathers Christmas: 168 ml/56 ml = 3, thus 3 × 6 ml = 18 ml), the 'rule of three' (i.e. via the amount of paint needed to paint 1 cm of the Father Christmas: 6 ml of paint for 56 cm, thus 0.107 ml for 1 cm, thus 0.107 × 168 cm = 18 ml for 168 cm) and the external ratio strategy (i.e. using the ratio between the amount of paint and the height in one Father Christmas: 6 ml/56 cm = 0.107 ml/cm, thus 0.107 ml/cm × 168 cm = 18 ml).

Most students were very sure or quite sure they gave a correct answer. On the five-point scale, 20 students indicated to be 'certainly correct', 16 to be 'probably correct' and the remaining 4 to 'have no idea'. Remarkably, the two 15–16-year olds who gave the correct non-linear answer in Phase 1, indicated to be 'probably correct' and to 'have no idea', respectively. Typical justifications of the students who indicated 'certainly correct' (while actually having answered incorrectly) were: "It's an easy problem. I just used the three numbers and the formula, so it must be correct", "It's logical, the Father Christmas becomes three times bigger". Students who expressed doubts about the correctness of their answer gave rather superficial and general reasons for their doubts that did not address the correctness of the applied model (e.g., "I'm not completely sure because I haven't carefully read the problem", "Maybe I made a computational error. That can always happen", "That was the first thing that came in mind, but maybe I didn't use the correct procedure", "You never are absolutely sure", "Mathematics is not my cup of tea, so I am not sure that my answer will be correct"). The great self-confidence observed in most students seems to indicate that for them, the linear model was self-evident. Moreover, students' reasons for being uncertain about the correctness of their given answer show their habits and beliefs when approaching word problems: The chance for success is mainly attributed to general mathematical ability, problems are read superficially, while possible mistakes are purely attributed to technical calculation errors.

Although students were quite sure about the correctness of their answer, they had great difficulties explaining *why* their method was the correct one. Initially, most students were unable to give any explanation at all. After insisting on a justification, students (1) referred to the fact that their solution is the most logical one, (2) explained that the Father Christmas is higher, so you need more paint and because it is three times as high, you need three times as much paint, (3) referred to the fact that the problem is about ratio or proportion. These superficial answers seem to indicate that students typically use the linear model in a spontaneous and thoughtless way and do not check whether this model is applicable in a given situation. Students do not seem to have clear arguments justifying its use, nor do they realise that there are competing models. Even the few students who realised that the enlargement acts in two dimensions did not necessarily give up the linear model. On the contrary, it was among these latter students that we observed the purest and most general expressions of the linear misconception ("It's 3 times bigger, not only the height but also the width. You can see it on the drawing. The whole thing is enlarged by factor 3, so you will need 3 times as much paint", "I knew it was enlarged, but not how much, so I calculated 168 : 56 and then I knew the multiplier", "Because the picture becomes larger, you need more paint, so you have to multiply by three", "It has the same shape, but it is enlarged, so you have to multiply the amount of paint by the same number"). Arguably, these students identified 'increase' with 'proportional increase', without making a clear mental representation of the problem situation.

3.4 Phase 2: Reactions to the weak form of cognitive conflict

After being confronted with the fictitious frequency table with the correct answer and the erroneous linear answer, only two of the 38 students involved in Phase 2 began to think about a mathematical model that also takes into account the width of the enlarged Father Christmas. Both students decided in favour of the correct, non-linear solution, although one of them argued that the width surely enlarges, but that it is impossible to know exactly how much. Being asked why they had given a wrong linear answer during Phase 1, both students admitted they had not thought about that (non-linear) solution and had not paid attention to the given drawings.

For the majority of the 36 students who stuck to their original linear answer, the confrontation with the frequency table with the answers of fictitious peers really induced a cognitive conflict too. They started wondering where that other frequently-chosen answer could come from. Remarkably, 13 students discovered that this answer was obtained by a multiplication by 3^2 (or 9), but most of them immediately rejected this method as erroneous ("They multiplied by three two times! You see, they made a mistake. My answer is correct"). The 23 other students who did not find the origin of the alternative answer searched for the rationale behind it in a more superficial way. For instance, they tried out 'randomly' some combinations with the basic arithmetical operations (+, –, ×, :) on the given numbers, regardless of their contextual meaning (e.g. trying to arrive at the alternative solution by subtracting the heights of the Fathers Christmas). The reasons of the 36 students for persisting in their linear solution were typically very general and extrinsic (e.g., "In general, I'm good in mathematics", "You better always stay with your first solution"). Besides, they often indicated that the linear answer was self-evident (e.g., "I would think my solution is much more logical", "It's evident, you cannot do it otherwise"), while the non-linear answer was qualified as 'counterintuitive' or 'illogical' (e.g., "In my opinion, this is a strange reasoning", "That's too far-fetched", "54 ml is quite a lot").

3.5 Phase 3: Reactions to the argumentation of a fictitious peer

In the third phase, another 14 students of the remaining 36 students (equally divided among both age groups) changed from their incorrect answer to the correct one. Apparently, the argumentation of the fictitious peer who answered the problem correctly provided them the insight that to maintain the same shape, a figure has to be enlarged in both dimensions, having a quadratic effect on the area of the figure. The 14 students who changed their answer were asked to explain why they originally gave the wrong linear solution. A first category of explanations referred to the fact that they did not approach the problem in a thoughtful way, but instead immediately and routinely (a student called it 'instinctively') started to reason proportionally ("I started to solve the problem that way, and, so to speak, I closed myself for the other reasoning. No other reasoning could come up in my mind any more, also because this is the easiest way"). Second, some students argued that they did not notice the width, because they were fixating on the problem statement which only referred to the height ("I only paid attention to the text … a little bit to the drawings, but above all to the text … and in the text only the height is mentioned", "In the text, only the height is given. If both the height and the width were mentioned, I probably would have used another formula"). Third, there were students who realised that the enlargement also had an effect on the width, but *deliberately* did not take it into account *because* the width was not explicitly mentioned in the problem statement ("I thought that the width was relevant too, but because in the text no reference was made to the width, I decided to work with the height only").

The justifications of the 22 students who decided to stick to their original answer after the confrontation with the reasoning process behind the correct non-linear answer, were diverse, but can be grouped into three different categories.

A first group of students gave the non-linear reasoning serious thought and they were torn between the two alternative solutions, but they finally opted for the familiar linear model because they insufficiently understood the mathematical principles relevant to this problem. Some realised that the enlargement had an impact on the width of the Father Christmasses too, but were unsure about how much this width increased ("Height and width are not that much related to each other", "The width changes too, but you cannot know how much"). Others struggled with the quadratic impact of a linear enlargement on a figure's area (or on the amount of paint, the indirect measure that is linearly related to this area) ("6 ml is for the *whole* Father Christmas, not only for the height. And 18 ml is for the *whole* large Father Christmas, for the height as well as for the width").

For a second group of students the argumentation of the fictitious peer was an immediate cause to formulate their mistaken linear belief more clearly and more convincingly than ever before ("if you need 6 ml for this area and this area fits three times in that area, you need three times more paint"). This reasoning is similar to that of those students in the first group who struggled with the quadratic impact of a linear enlargement on a figure's area in the sense that they both improperly assumed a linear relationship between two quantities. However, this second group differs in the sense that its linear reasoning seemed not to be inspired by any specific mental representation of the problem situation, but rather by an application of linearity 'anywhere'.

A third group of students justified their answer by referring to the implicit rules for solving school mathematics word problems (Lave, 1992; Verschaffel et al., 2000; Wyndhamn & Säljö, 1997). These students demonstrated a simplistic view on school word problem solving, assuming that all word problems can be solved by using simple mathematical calculations on the numbers given in the problem, and that real-world knowledge and context-based considerations should not be involved in the solution process. Sometimes, this type of argument also occurred in students belonging to the first two categories. Examples of justifications in this group are: "I think that the pupils who gave this answer make it too complex for a word problem", "In your calculation you can only involve numbers that are given", "The word problem says nothing about the width at all, so it must be wrong". Typically, these students rejected conflicting evidence arising from the given drawings ("In the drawing it is wider, but not in the word problem. The word problem is about the height only").

3.6 Phase 4: Reactions to the solution process of a fictitious peer

After the confrontation with the fictitious peer's solution process, namely actually drawing rectangles around the Fathers Christmas' irregular shapes, in Phase 4, another nine of the remaining 22 students (five 12–13- and four 15–16-year olds) exchanged their initial linear answer for the correct non-linear one. It appeared that for these students, the circumscribed rectangles seemed to function as a real 'Gestaltwechsel'[28] (Wertheimer, 1945), since they immediately and conclusively made a shift in their answers: "Oh yes, now I see it. Indeed, it is nine times larger because the small rectangle also fits nine times in the large one. With the help of these rectangles I understand it. I am sure now, it should be 54 ml". The uncertainty about *how much* the width increased, which was noted down several times in the previous phases, disappeared completely in these students. Before concluding the interview, these students were also asked why they originally gave a wrong linear answer and why they stuck so long to it. Their reactions were very similar to those given on this question by the students who found the correct response in Phase 3.

In the 13 students who stayed with their original linear answer during Phase 4, no reflections were made about a mathematical model that takes into account the increased width of Father Christmas. Most students justified their answer by expressing their beliefs about how to solve mathematical word problems ("Don't look too far for the solution of a school word problem", "It's possible, but the width is not mentioned in the problem statement") and about the role of, and the relationship between, textual and graphical information in a word problem ("It is nine times for the drawings, but three times for the word problem", "Drawings are less accurate", "You never should ground a solution in mathematics on a drawing. You have to ground it on formulas") (cf. also Phase 3). Some students could not give any justification at all or just repeated their mistaken linear belief, sounding like an attempt to convince the interviewer ("The little Father Christmas fits three times into the large one. It's the same for the rectangles. Three times more area, thus three times more paint").

Remarkably, the confrontation with the rectangles brought two of the 13 students who stayed with their original linear answer to express for the first time a strange misconception: The increase of area is different for the enlarged rectangles than for the irregular figures inscribed in these rectangles (in this case, the area scale factor was supposed to be *nine* for the rectangles, but *three* for the Fathers Christmas). A student formulated it this way: "What they do with the rectangle is correct – it enlarges in two directions. But within the rectangles is an irregular figure. And that's completely different. See here and here" (points at the white parts in the rectangles).

[28] The exchange of one perceptual viewpoint or perspective by another

3.7 Phase 5: Reactions to the explicit link with area measurement

In this final phase, five more of the remaining 13 students (three 12–13- and two 15–16-year olds) abandoned the wrong linear answer and moved to the correct non-linear answer. All five students correctly calculated the area of both rectangles and, to their astonishment, found that the area of the larger rectangle was indeed nine times the area of the smaller one. Three of them felt no difficulty in generalising this last finding to the 'irregular' Father Christmasses within these rectangles and to the amounts of paint needed to paint them. However, two students only succeeded in drawing this conclusion after the interviewer provided them with still an extra scaffold (e.g. after the interviewer had asked them to compare the amounts of paint needed to paint the two *rectangles*). Moreover, an ultimate choice for the correct answer did not necessarily remove all doubts (e.g., "OK, it is nine times more paint. But I still don't see why my original calculation was not correct").

Even after four types of increasing cognitive conflicts, eight students maintained their linear answer until the very end of the interview. While arguing their choice, these students stuck up even more strongly for their beliefs about how to solve word problems and about the role of drawings in word problem solving ("For the drawings, it is nine times, but this is not that relevant. In a word problem you are expected to work with the data that are provided in the text. Using drawings and measuring is less accurate") (see also Phases 3 and 4). Some of these persisting students also seemed to be worried about the role of 'amount of paint' as an indirect measure for 'area' in this context ("It's about 'consumption of paint' and you don't have to solve it via area", "For the area, it is nine times, but for the amount of paint, I'm not so sure. Millilitre is not referring to area, but rather to volume").

4. DISCUSSION

The interviews confirmed, once again, the existence of a very strong and deep-rooted tendency among 12–13- and 15–16-year olds to stick to the linear model when doing non-linear word problems about the enlargement of two-dimensional figures, even when confronted with very strong contradictory evidence for the tenability of that model in the given context. Indeed, after the first confrontation with the word problem, almost all students improperly applied the linear model and, in each of the consequent phases of the interview, only a limited number of students – often hesitatingly – gave up that model. After four types of increasing cognitive conflicts, providing increasingly strong evidence for a non-linear approach, still one fifth of the students stuck to the linear model.

More importantly, the interview study provided a lot of information about the reasoning and problem-solving processes of students falling into the 'proportionality trap' and the mechanisms behind it. The results enabled us to identify the role of different aspects of students' knowledge base that were responsible for their inappropriate proportional responses. These aspects can be grouped in three distinct categories, which we will describe and discuss in detail now. However, it is important to mention here that students generally cannot be put into one of these categories. Most often, the reactions of a student during the five phases of the interview involved a complex interplay of elements belonging to different categories with some of these elements being more prominent in particular interview phases.

A first category of explanations refers to the *intuitiveness of linear relationships*. According to Fischbein (1987), intuitive cognitions have an obvious, self-evident, and coercive character, receive great confidence and persist despite formal learning. These characteristics seem to apply to the incorrect reasonings of the students in our study too, particularly in the first phase of the interview: The use of linear relationships was perceived as correct without a need for any further justification, students were overconfident in it, and were reluctant to question the correctness of their linear approach when confronted with conflicting evidence. Proportions appeared to be deeply rooted in students' intuitive knowledge and were used in a spontaneous or even unconscious way, which made the linear approach quite natural, unquestionable and, to a certain extent, inaccessible for introspection or reflection. While thinking aloud, most students immediately used proportions, they were convinced about the appropriateness of the proportional model and of the correctness of their answer, but it was virtually impossible for them to justify what they did. Later on during the interview, some students qualified the non-proportional solution as 'counterintuitive' or 'illogical' (see, e.g., students' reasons for persisting in their linear solution in Phase 2). There also seems to be a parallel between students' problem-solving process – especially in the very first encounter with the problem – and the 'intuitive rules' described and studied by Tirosh and Stavy (1999a, 1999b). These authors claim that there are some common, intuitive rules that come in action when students solve problems in mathematics and sciences. Two such rules are manifested in comparison tasks: 'more *a*–more *b*' and 'same *a*–same *b*'. In the problem we presented to our students, it is quite natural (and correct) to apply the 'more *a*–more *b*' rule (the *more* height, the *more* area/paint). However, the 'same *a*–same *b*' reasoning might occur too (figures share the *same* shape, so everything enlarges by the *same* factor), leading to an incorrect '*k* times *a*–*k* times *b*' judgement (three times more height, so three times more area/paint). Expressions in line with these schemes occurred several times during the interview and were sometimes phrased literally.

Second, besides the intuitiveness of linear relationships, the observations during the interviews clearly show that students often have *inadequate habits and beliefs* about solving word problems in a school context, or about mathematical modelling in general, a conclusion that is supported by a vast amount of research (see, e.g., Verschaffel et al., 2000; Wyndhamn & Säljö, 1997). As several authors have stressed, mature mathematical modelling involves a complex, cyclical process consisting of a number of subsequent steps: understanding the situation described; selecting the elements and relations in this situation that are relevant; building a mathematical model and working through it; interpreting the outcome of the computational work in terms of the practical situation; and evaluating the results and the applied model itself (Burkhardt, 1994; Greer, 1997; Verschaffel et al., 2000). In the modelling process observed in many of our students, some of these steps were completely bypassed. Little effort was invested in understanding the problem situation and in making a clear mental representation of the relevant elements and relations. The mathematical model then mainly occurred on the basis of 'reflex-like recognising', and was almost immediately translated into calculations. These calculations received most time and attention in the problem-solving process. The superficial character of students' modelling process also appeared in the last phases of the modelling cycle. No critical evaluation of the model itself, or of the results obtained by applying this model, was undertaken. With the exception of a quick control on calculation errors, students did not spontaneously verify their answer by means of their common-sense knowledge or the given drawings, and in no phase of the modelling cycle did students spontaneously compare the applied linear model with alternative models and even when confronted with these alternatives, they were not seriously taken into consideration. So, in many cases students' improper use of linear reasoning can clearly be seen as a symptom of an immature and even distorted disposition towards mathematical modelling.

Whereas the two previous explanatory categories were quite general (in the sense that they are not specifically related to students' belief that if a figure enlarges k times, the area and/or volume enlarge k times too), the third category of explanations relates precisely to these content-specific issues. A first content-specific element coming forward from this in-depth study is that many students (younger as well as older ones) suffer from *shortcomings in their geometrical knowledge*, especially about the effect of a similarity on the lengths and area of a figure. For example, students confused area and volume or did not recognise indirect measures for area (such as the amount of paint)[29]. Other utterances of this fuzzy geometrical knowledge are the convictions that (1) when a figure is enlarged but maintains its shape, the height and width do not necessarily increase by the same factor, (2) enlargements have a different effect on the area of a regular figure than on the area of an irregular one, and (3) only regular figures have an area. Although students learned these basic concepts and principles of geometry and measurement at the elementary school level, they seemed to have a bad or weak understanding of them, or at least they were not able to apply them correctly. As such, this fuzzy geometrical knowledge did not really *cause* students to apply a linear method to the non-linear word problem, but it did prevent them from discovering the correct solution and from unmasking the inadequacies in their own linear solution. There is a second content-specific element, however, that probably *did* lead students towards the linear solution. When enlarging a geometrical figure (like in the word problem offered to the students) while maintaining the same shape, the figure is enlarged in a *linear* way, i.e., the figure becomes k times higher *and* moreover k times wider. When students approach such a problem, their thinking may exactly be pointed towards the linear changes in the situation. In some cases, in our study, students' improper use of linear reasoning seems to be the result of a conscious and deliberate application of linear functions. These students really were convinced that the linear model was applicable in that situation. In their conviction, the same scale factor (namely factor 3) applied for both the lengths and the area of a geometrical figure. A student stating "The height and the width of the figure are tripled, so the area is tripled too" clearly has an adequate mental representation of the problem situation (since the problem statement only mentioned the height), and he was not purely focussing on superficial characteristics of the problem situation. It was precisely this mental representation that *convinced* him of the correctness of his linear solution. Also in Follow-up study 3 in chapter 3, we observed a similar effect, namely to draw a reduced shape, students divided both dimensions of the given shape by the linear scale factor, but, as a result, they made more linear errors!

[29] Note that Follow-up study 4 in chapter 3 specifically aimed at verifying this kind of difficulties and their influence on students' overuse of linearity. In that study, we found no significant impact of the use of direct or indirect measures, but apparently it does play a role for some students.

Chapter 5

A TEACHING EXPERIMENT

1. INTRODUCTION

The studies presented in chapter 2 revealed that many secondary school students improperly apply linear models when solving application problems about the effect of a linear enlargement or reduction of a geometrical figure on the perimeter, area, or volume of that figure. In chapter 3 we reported several studies showing that even with considerable support, such as self-made and ready-made drawings or metacognitive hints, only very few students made the shift from incorrect linear to correct non-linear reasoning. The interview study presented in chapter 4 yielded a more fine-grained picture of the thinking process underlying students' improper linear reasoning and how this process is affected by different aspects of their knowledge base. The next stage of our research program – which is the focus of the present chapter – involved the design, implementation, and evaluation of an experimental learning environment aimed at overcoming students' overuse of linearity in the context of the enlargement/reduction of geometrical figures, and at developing in students a deep conceptual understanding of proportional and non-proportional relations and situations in this domain (see also Van Dooren, De Bock, Hessels, Janssens, & Verschaffel, 2004).

2. DESIGNING A LEARNING ENVIRONMENT

The goal of the current teaching experiment was to break students' deep-rooted tendency to apply linear strategies in non-linear situations, more specifically in the context of the relationship between the linear measures of a figure and its perimeter, area, and volume. For this goal, a series of 10 one-hour experimental lessons – including all teacher and learner materials – was developed for use with 8th graders.

We first present an overview of the contents that were treated in these lessons, and illustrate this overview with some key learning tasks and exercises. Afterwards, we discuss the more general principles underlying the design of the lessons, and refer back to some of the key activities presented here.

2.1 Overview of the lesson series

Lesson 1 and 2 focused on the concept of similarity itself. Students learned to recognize and construct similarly enlarged and reduced figures and objects and understand their properties. A key activity was built around the famous painting 'Golconde' by the Belgian surrealist René Magritte (see Figure 5-1). Together with the 'original' painting, different variants were given and students were asked to select the appropriate one, and describe how to find out which variant is appropriate. Rectangular objects are particularly useful to introduce the topic of similarity. A 'stretched' rectangle is still rectangular, so to judge similarity, students explicitly need to focus on height/width ratios (whereas similarity can be 'holistically' judged for, e.g., squares or circles). On the other hand, the painting *within* the rectangle addressed students' informal knowledge of what looks the same and what does not. Subsequent activities included examining similarity of diverse drawings and of two- and three-dimensional objects (cans, boxes, envelopes, bottles, ...), constructing similar enlargements and reductions of given figures, and finding 'easy' strategies to construct similar figures. Afterwards, a whole-class discussion was organised, leading to:

- the properties of similar (two-dimensional) figures and (three-dimensional) objects,
- methods of constructing similar figures in different sizes,
- the meaning of the enlargement or reduction factor k, ways to find k and its role in constructing similar enlargements or reductions.

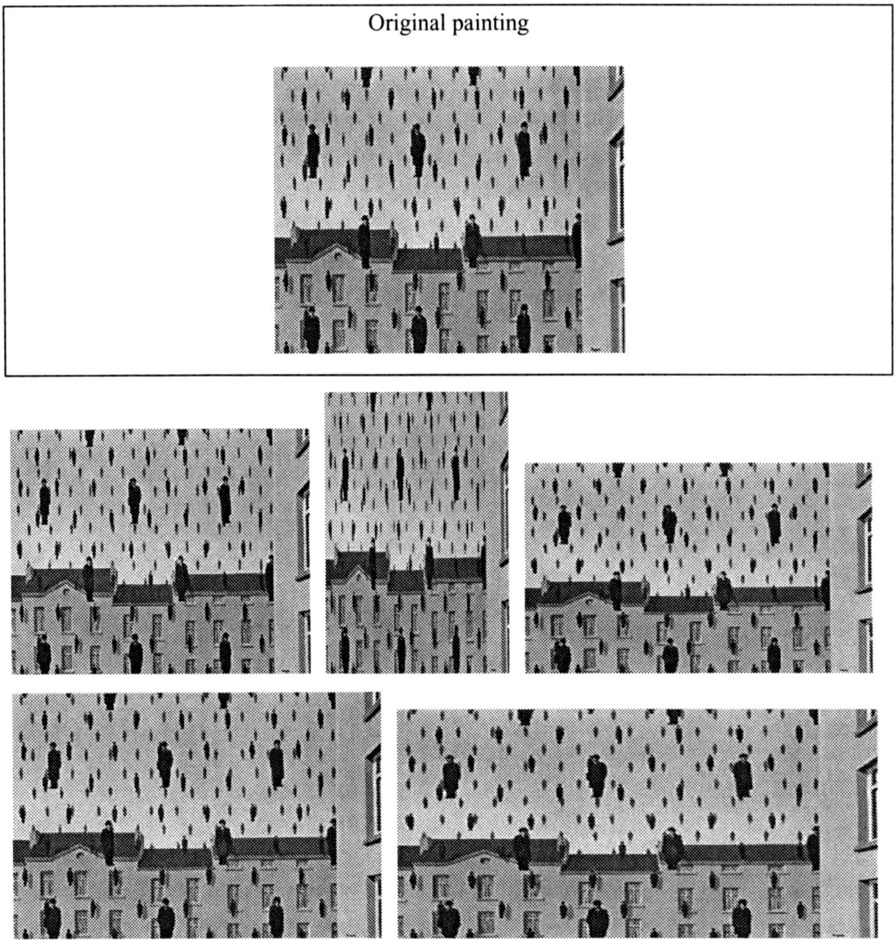

Figure 5-1. Painting by Magritte and five variants

The linear relations in similar figures were the central topic in lesson 3. The goal was to make students' understanding of the proportional relations explicit, to show how this knowledge applies to similar figures and how it can be used to predict lengths in other sizes of the figure. In one task, a picture of The Simpsons was provided in four different sizes, and students were asked to complete two tables about these picture sizes and then to put some of these relationships in a graph. Afterwards, a group discussion was organised leading to conclusions such as:

- each length in figure A is *k* times that length in figure B (and the same holds for any other pair of figures),
- the ratio of any two corresponding lengths is the same within each figure,

- predictions about lengths in other sizes of the figure can be made by interpolation or extrapolation of the linear graph.

In lessons 4 and 5, the ideas of linear growth of perimeter and the quadratic growth of area were introduced and explored. We pursued students' understanding and ability to apply the principle that if a figure is enlarged or reduced k times, the perimeter of that figure enlarges k times too, but the area enlarges k^2 times. The proportional growth of perimeter in enlargements was introduced first by a task on cola bottles. A 1.5 litre and a 0.5 litre bottle were shown (see Figure 5-2) and students were asked the following questions:

Figure 5-2. Two different sizes of cola bottles

- Is a big cola bottle of 1.5 litres similar to a small cola bottle of 0.5 litres? (*The conclusion was that they are similar if one ignores the bottleneck)*
- If you strip the labels of the bottles, the label of the small bottle is 5 cm high by 20 cm wide. The label of the big bottle is 7.3 cm high. Considering that both bottles are similar to each other, what should be the width of the label of the big bottle? Check if this is true by measuring on the big bottle.
- If you know the perimeter of the large bottle, can you calculate the perimeter of the small one? Again, check by measuring the bottles.

This introductory task was followed by work in small groups on quadratic growth of area in enlargements. Students were confronted with the following pancake problem:

- It's Anne's birthday and her mother is going to make pancakes, using three pans of different sizes. Anne asks her friend: "You may choose between two big pancakes (30 cm diameter), four regular ones (20 cm diameter) or six small ones (15 cm diameter)."
- Her friend reasons as follows: "You better choose six small pancakes because 2 × 30 cm = 60 cm, 4 × 20 cm = 80 cm and 6 × 15 cm = 90 cm"

- What do you think about the reasoning of Anne's friend? Compare your solutions.
- How many of the smallest pancakes do you need to have the same amount as one large pancake? Why?

The task was meant to elicit discussion by introducing a cognitive conflict. Students worked in groups, and some were inclined to calculate areas of the different pancakes by applying the learnt formulae, while others found the reasoning by Anne's friend intuitively acceptable and appealing. In a classroom discussion that followed the group work, the different group solutions were compared and various approaches to the problem were explored, e.g. calculating areas, using cardboard discs as models (weighing the cardboard discs, cutting and covering a large disc with pieces of a smaller disc, …). Afterwards, the effects of enlargements and reductions on the perimeter and area of a plane figure were investigated in a whole-class discussion by means of multiple representations (formula, table, drawing, and graph) and for a diversity of figures (squares, circles, rectangles, and irregular figures). In Figure 5-3 we illustrate this approach for a square.

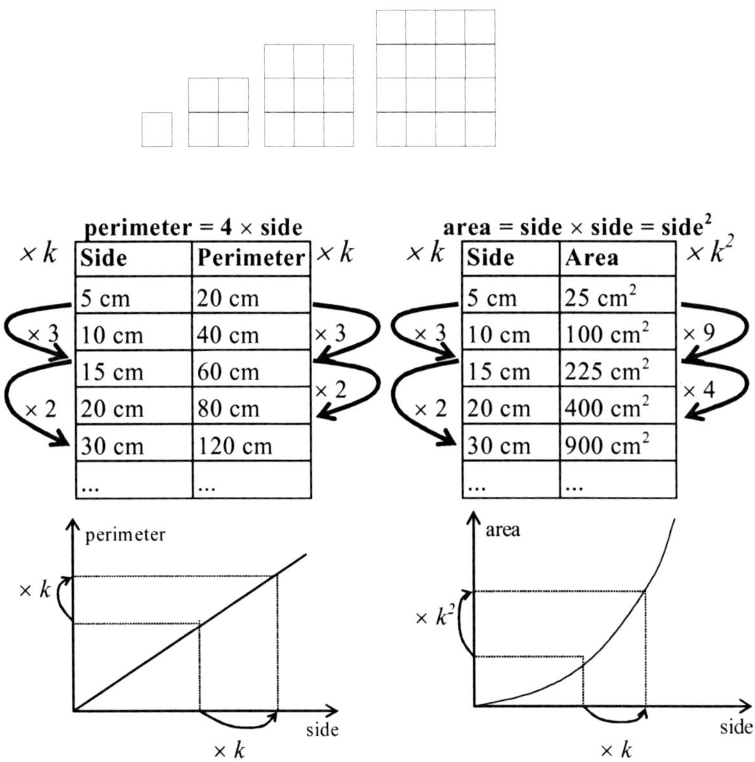

Side	Perimeter
5 cm	20 cm
10 cm	40 cm
15 cm	60 cm
20 cm	80 cm
30 cm	120 cm
…	…

Side	Area
5 cm	25 cm^2
10 cm	100 cm^2
15 cm	225 cm^2
20 cm	400 cm^2
30 cm	900 cm^2
…	…

Figure 5-3. Different representations of an enlargement of a square

At the end of lessons 4 and 5, several applications were made, individually or in dyads, e.g.: "Tiny's parents bought 40 m^2 of floor tiles. They needed 30 m^2 for the living room. Tiny's bedroom is only half as long and half as wide. Are there enough tiles left to put a new floor in her bedroom?"

The central topic of lessons 6 and 7 was the cubic growth of volume. We pursued students understanding of and ability to apply the principle that if a solid is linearly enlarged or reduced k times, the volume of that solid enlarges k^3 times (and its surface area enlarges k^2 times). In small groups, students first worked on the doghouse problem: Tommy wants to make a doghouse, following the model given in Figure 5-4.

- What's the area of all components? How much wood should he approximately order?
- His uncle has a much larger dog. If he were to construct exactly the same doghouse, but twice as large, how much wood would he need?
- Calculate the volume of the small doghouse, and then estimate the volume of the large doghouse. How much larger is it? (Try to explain using a simpler model like a cube).

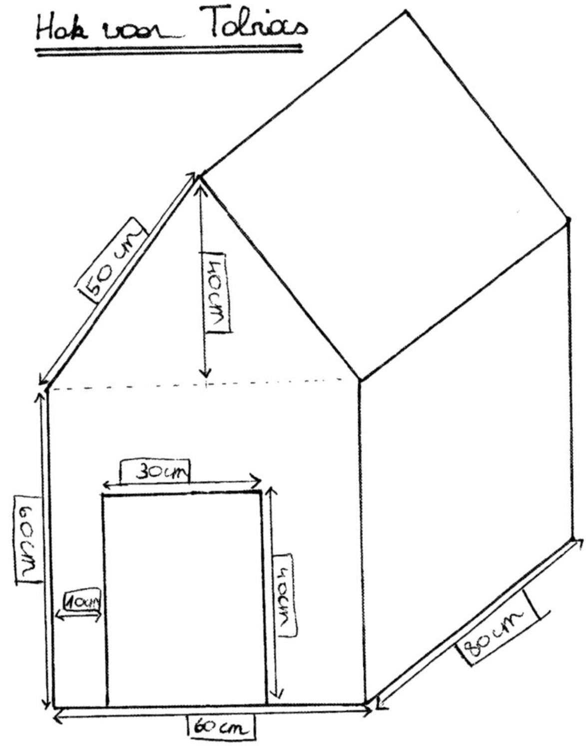

Figure 5-4. A house for Tobias

This activity was completed by a classroom discussion in which the solutions from the different groups were commented upon and possible approaches and solutions were explored. Afterwards, students investigated the effect of enlargements and reductions of solids on their area and volume in a whole-class discussion again by means of multiple representations (formula, table, drawing, and graph) for a diversity of solids (cubes, rectangular prisms, cylinders, spheres, …). This approach is illustrated for a cylinder in Figure 5-5.

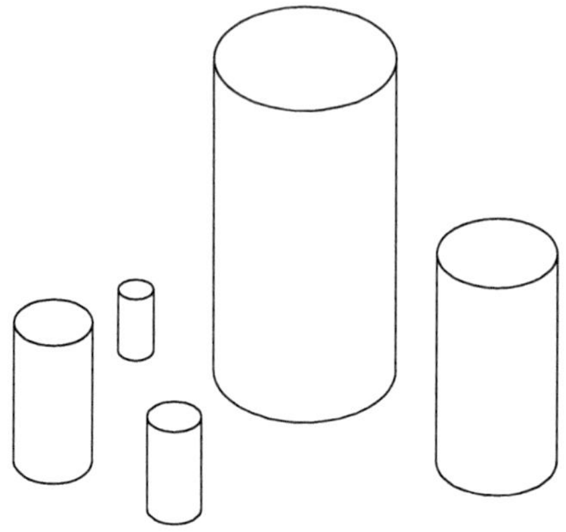

Area = 2 × (π r^2) + (2 π r) × height

Radius	Height	Area
1cm	8 cm	55.8 cm^2
1.5 cm	12 cm	125.6 cm^2
2 cm	16 cm	223.2 cm^2
3 cm	24 cm	502.2 cm^2
…	…	…
6 cm	48 cm	2008.8 cm^2
…	…	…

× k; × 2; × 3; × k^2; × 4; × 9

Volume = π × radius × radius × height

Radius	Height	Volume
1cm	8 cm	24.8 cm^3
1.5 cm	12 cm	83.7 cm^3
2 cm	16 cm	198.4 cm^3
3 cm	24 cm	669.6 cm^3
…	…	…
6 cm	48 cm	5356.8 cm^3
…	…	…

× k; × 2; × 3; × k^3; × 8; × 27

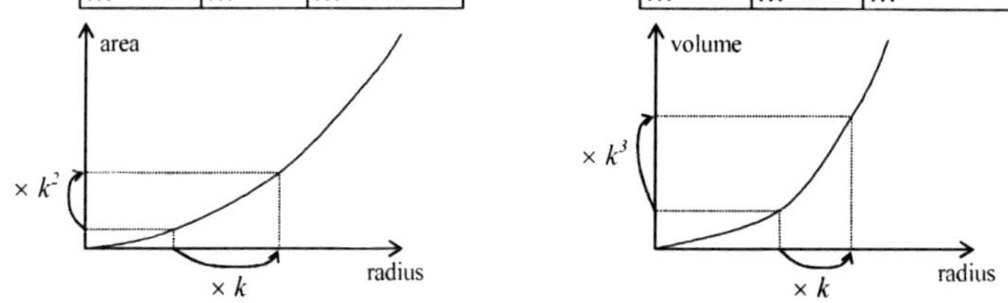

Figure 5-5. Different representations of an enlargment of a cylinder

Afterwards, several applications were made, individually or in dyads, e.g.:

- An apple grower sells two sizes of apples. The first one has an average diameter of 6 cm and costs 10 eurocent and the second one has an average diameter of 9 cm and costs 20 eurocent. Which apple size is the more economical to make apple sauce?
- Another farmer sells the same apples at other prices. Now, the 6 cm apples cost € 1.00 for a kilogram, and the 9 cm apples cost € 1.20 for a kilogram. Which apple size is the more economical now?

If necessary, the teacher gave hints such as "Compare both sizes of the apples. What is the enlargement factor (*k*)?" or "The apples have a similar shape. How much more does a big apple weigh compared to a small apple?"

The teaching experiment concluded in lessons 8 to 10 with an integrative project on the life and work of the gnomes (Poortvliet & Huygen, 1977). The students were first confronted with the challenging task of examining a claim (made in the book) that a gnome 15 cm high weighs about 300 grams, assuming that a gnome is similar to a human being. Applying the taught principles about the change of volume in a reduced object and their real-life knowledge about the size and weight of a human, the students discovered that a more appropriate estimation would be 70 000 grams / 12^3 = 40.6 grams, or conversely, that if the estimation in the book was correct, a human would weigh 300 × 12^3 = 518 400 grams or more than 500 kilos.

Thereafter, students had to solve different questions about the world of gnomes (assuming that a gnome is similar to a human being, but 12 times smaller), some relating to perimeter, others to area or volume. For many problems, students needed to make estimations based on their real-life knowledge or to collect additional information from additional sources. Examples of such questions were:

- How long is the belt of a gnome?
- What is the area of the sole of a gnome shoe?
- How much coffee is there in a cup for gnomes?
- How much fabric do you need to make a skirt for a woman gnome?

2.2 General design principles

With respect to the more general purposes and design principles embedded in the lesson series, the results and conclusions of the studies described in chapters 2 to 4 were taken into account. Throughout the lesson series, we explicitly addressed shortcomings in students' geometrical knowledge base, such as the concept of area in relation to irregular figures, the principles governing an enlargement or reduction of a figure (lessons 1 and 2), and strategies to determine perimeters, areas, and volumes. The issue of direct and indirect measures for area and volume was problematised and we brought variation in (the formulation of) proportional and non-proportional tasks. We also attempted to break students' intuitive use of linearity by thoroughly discussing the applicability of a (proportional) model for a given situation. We specifically incorporated an exploration of the capacity of linear functions to grasp many relations in similarly enlarged or reduced figures (see, e.g., the task in lesson 3), but also demonstrated their limitations in modelling certain other situations. For example, the 'pancakes' problem in lessons 4 and 5 was designed to elicit a cognitive conflict in students and meaningful discussions during small group work and classroom discussions. Finally, we aimed at addressing students' inappropriate inadequate habits and beliefs during mathematical word problem solving, such as the idea that mathematical problems can be solved through the application of one or two simple operations, that all required data are given in a problem, that every problem has an exact numerical answer, and that drawings are not helpful for approaching mathematical problems (see for example, the 'doghouse' problem in lessons 6 and 7 or the integrative project lessons 8 to 10).

At a more general level, the development of our learning environment was inspired by design principles from the literature on realistic mathematics education (de Lange, 1987; Gravemeijer, 1994; Treffers, 1987), powerful learning environments, conceptual change and the enhancement of higher-order thinking skills (Collins, Brown, & Newman, 1989; De Corte, Verschaffel, & Masui, 2004; Vosniadou, Ioannides, Dimitrakopoulou, & Papademetriou, 2001). First, the lessons were interspersed with a rich variety of meaningful, realistic, and attractive problem situations, aimed at challenging particular mathematical (mis)conceptions and/or stereotyped and superficial modelling behaviour. As an illustration, we refer to the integrative project that the students worked on during the lessons 8 to 10. These lessons were devoted to a project (to be realized in small groups) about the book 'Gnomes' (Poortvliet & Huygen, 1977), describing in a very detailed and touchingly realistic way all aspects of the life of the gnomes. During the project, the students had the opportunity to apply all newly acquired mathematical insights in complex, attractive and challenging problems situated in the world of gnomes, where all lengths are 12 times smaller. Second, the learning environment relied on a combination of instructional techniques (e.g. small-group work, changing roles of group members, exchanges between groups, whole-class discussions, etc.) that have proven to be successful in enhancing students' deep understanding and higher-order thinking skills (e.g., articulation and reflection). A third characteristic was that multiple representations of the learning contents (such as drawings, schemes, tables, graphs, formulas, and words) were used and their reciprocal relationships were accentuated to enhance deep-level learning (see, e.g., Ainsworth, Bibby, & Wood, 2002; National Council of Teachers of Mathematics, 1989). For example, as illustrated in Figure 5-3, the quadratic relationship between the side and the area of a square was analysed in the following representations:

- the formula '$A = side \times side = side^2$' for the area of a square (in which a quadratic factor can be observed, and in which *side* can be algebraically substituted by $1.5 \times side$, $2 \times side$, $3 \times side$ etc., or by different numerical values),
- a table in which the area is calculated for several values of the side (and in which the areas for different sizes of sides can be compared),

- a graph representing the relation between the side and area (and which illustrates the non-linear character of the relationship),
- a drawing (in which a large square can be 'covered' with small squares to get a visual comparison of their areas).

3. RESEARCH METHOD

In this study, 35 8th-graders (aged 13-14) were involved. They belonged to two intact classes from two secondary schools in Flanders, attracting comparable student populations. Both groups consisted of about equal numbers of boys and girls. One class (of 18 students) was assigned to the experimental condition and followed the experimental lesson series. All lessons in this group were videotaped and all student notes were collected. The other group (of 17 students) acted as a control group. These students followed the regular lessons, in which none of the contents under consideration was treated. Prior to the teaching experiment, the mathematics teacher in this group was not informed about the topic of the study.

Learning gains in both groups were assessed by means of a word-problem test consisting of 2 proportional items (about the perimeter of an enlarged square and of a circle), 4 non-proportional items (about the area and the volume of an enlarged square/cube and of an irregular figure) and 2 unrelated buffer items. Table 5-1 gives an example of a proportional and a non-proportional item. Three parallel versions of this test were constructed. Each student of the experimental group solved one version of the test before the intervention (pretest), another version immediately after the intervention (post-test), and a third version three months afterwards (retention test). Students in the control group received only a pretest and a retention test.

Table 5-1. Examples of word problems used in the test

Proportional item (perimeter)	Non-proportional item (volume)
Steve needs 10 minutes to dig a ditch around a square sandcastle with a side of 50 cm. How much time will he approximately need to dig a ditch around a square sandcastle with a side of 150 cm?	In his toy box, John has dice in several sizes. The smallest one has a side of 10 mm and weighs 800 mg. What would be the weight of the largest die (with a side of 30 mm)?

All problems were open-ended questions, and students had to write down their answer and their calculations. Answers were scored either as correct or as incorrect, and incorrect answers were further categorized as follows (based on an analysis of the students' written calculations): application of proportional methods to non-proportional items (P), application of non-proportional methods to proportional items (NP), other errors (O).

4. RESEARCH QUESTIONS AND HYPOTHESES

The goal of the current study was to test whether a learning environment with the above-mentioned characteristics could cause a substantial reduction in students' tendency to produce linear answers in situations where they are not correct. Based on the studies reported in chapters 2 to 4 we expected that on the pretest, experimental and control group students would generally respond correctly on the proportional items and incorrectly on the non-proportional items, because of their tendency to apply proportional strategies for these latter items too. Due to the learning experiences in the experimental lessons, we expected a significant progress in the performance of the experimental group on the post-test – more specifically on the non-proportional items – and that this progress would largely persist on the retention test. For the control group, no significant evolution from pretest to retention test was expected, because these students were not involved in any learning activities specifically addressing the errors under consideration.

5. RESULTS

A '2 × 2 × 3' repeated measures ANOVA was conducted with 'Group' (experimental vs. control), 'Proportionality' (proportional vs. non-proportional items) and 'Test' (pretest vs. post-test vs. retention test) as independent variables and students' number of 'Correct answers' on the word problems as the dependent variable. Based on our hypotheses, we expected a significant 'Group' × 'Proportionality' × 'Test' interaction effect, which was confirmed by the ANOVA[30]. An overview of the percentages of correct answers is given in Table 5-2.

[30] $F(1,488) = 4.80, p = 0.0290$

Table 5-2. Percentage correct answers and standard deviations of the experimental and control group on the proportional and non-proportional items at the three test moments (pretest, post-test and retention test)

	Proportional items						Non-proportional items					
	Pre		Post		Retention		Pre		Post		Retention	
	%	SD	%	SD	%	SD	%	SD	%	SD	%	SD
Experimental	83.3	7.8	58.3	7.8	52.8	7.8	29.2	6.2	61.1	6.2	50.0	6.2
Control	85.3	8.1			73.5	8.1	13.2	6.4			16.1	6.4

As expected, on the pretest there was a significant difference between the performance on the proportional items (which were solved very well) and the non-proportional items (which were mostly solved incorrectly due to linear reasoning), both in the experimental group and in the control group[31]. Again, this is evidence for students' tendency to produce linear answers in non-linear situations. Furthermore, at the pretest there was no significant difference between the experimental and control groups, indicating that both groups were indeed comparable. Both groups performed almost the same on the proportional items, and the difference for the non-proportional items was also not significant[32].

We will first discuss the results of the control group. Afterwards, we will contrast these results with those of the experimental group.

We did not expect a significant evolution from pretest to retention test in the control group. In line with this expectation, we observed only a very small, non-significant increase in the performance on the non-proportional items (from 13.2% to 16.1% correct answers) and a non-significant decrease in the number of correctly answered proportional items (from 85.3% to 73.5%) from pretest to retention test. A qualitative analysis of the protocols in this group showed that in about 80% of the cases students applied linear strategies to the non-linear items on the pretest and the retention test (P-errors). A small increase in the number of overgeneralisations of non-linear strategies to linear items (NP-errors) was observed (from about 11% to 18%) between pretest and retention test. Apparently, as an effect of retesting, a few students started to apply non-linear solution methods to the linear

[31] Experimental group: $t(488) = 6.56, p = 0.0001$; control group: $t(488) = 8.48, p = 0.0001$

[32] Despite the non-significant outcome of the Tukey test, both groups' performances on the non-proportional items seemed to differ considerably. An additional analysis of covariance (ANCOVA) was performed, predicting the performances on the non-proportional items on the retention test on the basis of the group (experimental/control) *correcting for the performances on these items on the pretest* (thus cancelling out any differences between the groups at the pretest). The corrected means of both groups at the post-test (44.3% and 22.3% for the experimental and control group respectively) were still statistically different, $F(1,32) = 6.42$, $p = 0.0164$.

problems they solved correctly before. This is similar to observations made in our earlier paper-and-pencil studies.

The experimental group significantly improved on the non-proportional items from pretest (29.2%) to post-test (61.1%)[33], followed by a non-significant decrease in the performances on the retention test (to 50.0% correct answers). This means that the experimental group made a significant progress on the non-proportional items, and that this progress persisted over several months. However, this improvement in performance was not as high as we had anticipated. Contrary to the results for the non-proportional items, the score of the experimental group on the proportional items decreased significantly from 83.3% correct answers on the pretest to 58.3% on the post-test[34], and went further down from post-test to retention test (although not significantly) to 52.8%. Apparently, in line with our earlier studies, when these students discovered that some problems can not be solved by applying proportions, they started applying non-proportional solution schemes to proportional problems too. A qualitative analysis of the answers of the experimental group revealed first of all that on the pretest about 70% of all the solutions on the non-proportional items could be characterized as proportional. This number of erroneous linear answers strongly decreased in the post-test to about 18%, while in the retention test, the percentage raised again to about 30%. But students who no longer applied linear solutions to solve non-linear problems during the post-test or the retention test, did not always perform better than in the pretest. In the post-test and retention test, many of them made errors in applying non-linear solutions on these non-linear problems (e.g., confusing area and volume, taking the square of a given number, …). The qualitative analysis also confirmed the overgeneralisation of non-linear strategies (NP-errors) after the experimental lessons. While on the pretest only 13% of all the solutions to linear items could be characterized as applications of non-linear strategies (NP-errors), this number increased to 36% on the post-test and on the retention test!

The evolution of the performance of the experimental group on the tests revealed the fragile and unsteady nature of these students' emerging non-proportional reasoning scheme. An analysis of the videotaped experimental lessons supports this conclusion. We restrict ourselves to the most important conclusions of the analysis of these fragments, in an attempt to explain the somewhat disappointing test results of the students from the experimental group.

[33] $t(488) = 3.09, p = 0.0001$

[34] $t(488) = -2.62, p = 0.0090$

First of all, certain lesson fragments revealed that non-linear relations and the effect of enlargements on area and volume remained intrinsically difficult and counterintuitive for many students. A striking example comes from a student raising the following question in the final lesson: "I really do understand now why the area of a square increases 9 times if the sides are tripled in length, since the enlargement of the area goes in two dimensions. But suddenly I start to wonder why this does *not* hold for the perimeter. The perimeter also increases in two directions, doesn't it?"

A second, related issue coming forward from the video analysis is that, for the students in our intervention study, proportional reasoning was already well established and practiced, so that they initially over-generalised proportionality to nearly all problems. But as soon as its applicability was questioned for some situations, the students again started to make hasty overgeneralisations. Some students went on switching back and forth between applying proportionality 'anywhere' and applying it 'nowhere' during the course of the experimental program, thereby relying only on superficial cues just as they did before. Whereas the instructional goal was that students would make fundamental changes in their conceptual framework to reflect and deliberately discriminate between the concepts and strategies that are relevant in a certain situation and those that are not, some students seemed to perceive the goal of our intervention as being that the old linear thinking should entirely disappear.

Third, the analysis of the videotaped lessons showed that while working in groups on the learning tasks, students very often did not engage in thoughtful, effortful processing as much as we wanted them to do. Instead, they often stuck to very superficial cues, prompting rapid, global strategy choices and evaluations. Contrary to our expectations, the experimental learning tasks – that were based on realistic and attractive learning tasks and that intended to induce cognitive conflict where possible – often failed to create a real feeling of 'personal relevance' (Limón, 2001) in the students. When cognitive conflicts arose during group work, local inconsistencies in students' thinking were often 'patched up' in a rather superficial way. And whereas the instructional methods of group work, exchanges of ideas and results between groups, and classroom discussions were expected to enhance metacognitive awareness, most of the time the discourse only shortly focused on the genuine exchange and grounded discussion of ideas. These exchanges were moreover often based on the 'authority' of the mathematically more able students. In sum, the realistic character of the learning tasks and the interactive and thought-provoking character of the instructional methods did not always reach their goal.

6. CONCLUSIONS AND DISCUSSION

In general, the results of this study confirmed our expectations. Initially, both the experimental and the control group performed well on the proportional items but often failed on the non-proportional items, due to the application of linear methods. After the experimental lesson series, the experimental group applied linear solution methods less often on the non-linear items on the test. However, a considerable part of the non-linear items on the post-test and retention test were still solved erroneously by this group, either due to linear reasoning or to errors in the application of non-linear strategies. Moreover, at the post-test and retention test the students of the experimental group began to make more errors on the proportional problems, because they over-generalised the newly learnt non-proportional strategies to the proportional problems they previously solved very well. Apparently, after the lessons, the students still experienced serious difficulties in determining which model to use in which situation. Therefore we can hardly argue that the lesson series has fully reached its goal. Many students did not develop a deeper understanding of (non-)proportionality and a disposition to distinguish between situations that can and cannot be modelled proportionally. The non-proportional reasoning scheme that emerged in some experimental students still remained very fragile and unstable.

To change the students' deep-rooted habits and beliefs contributing to a superficial modelling process, a teaching intervention during a longer time period or spread out over a longer time period seems necessary. The 10-hour experimental lesson series could deal with the core mathematical content knowledge that is necessary to appropriately solve problems about the change in perimeter, area, and volume of enlarged or reduced geometrical figures. But this intervention of only 10 hours, which was moreover not completely embedded in the regular mathematics curriculum and the regular classroom practice and culture, was clearly not sufficient to radically change students' habits and beliefs with respect to mathematical problem solving, a crucial factor in the occurrence of improper linear reasoning (see the interview study in chapter 4). Evidence on the possibility of changing students' inappropriate conceptions and beliefs with respect to mathematical modelling by immersing them during an extensive period in a new classroom culture can be found in Verschaffel and De Corte (1997).

Besides the difficulty of changing students' habits and beliefs in a limited time range and in a restricted instructional setting, we wonder whether such intervention should not take place earlier than in Grade 8. We actually opted for this age group since its regular mathematics curriculum corresponded most to the contents treated in our experimental lesson series. But since these 8th-graders already had so much opportunity to practice the proportional reasoning scheme and experienced usefulness of that scheme, this prior knowledge seems to continue interfering in their learning process. A recent study by Van Dooren et al. (2005, see also chapter 1) on students' solutions of arithmetic word problems confirmed that students' over-application of proportionality originates in parallel with students' growing proportional reasoning skills and the curricular attention this topic receives. In that study, the overuse of linearity proved to be already modestly present in the 2nd grade, increased up to Grade 5, before only slightly decreasing from Grade 6 to 8. Therefore, it might be more appropriate to intervene much earlier in order to prevent – rather than remedy – the overuse of linearity. At the first time when students meet proportional relationships in their mathematics curriculum in a more formal way (i.e. in 4th and 5th grade), they should also be confronted with counterexamples (of situations where linearity does not work), and learn to distinguish between situations that can be modelled proportionally and situations with another underlying mathematical structure. In this way, one can possibly avoid or counter early the intuitive presupposition that any relation between magnitudes is proportional. And when, at a later age, students discover, for example, that an area increases k^2 times if a figure is enlarged k times, they might experience less incongruency with their presuppositions and be more susceptible to generalizing and formalizing this principle, which was the goal of the current teaching experiment.

Chapter 6

STEPPING OUTSIDE THE CLASSROOM

1. INTRODUCTION

If there is one overall conclusion that we can deduce from the series of studies presented so far – from the investigations based on collective tests reported in chapters 2 and 3, over the in-depth interview study presented in chapter 4, to the teaching experiment reported in chapter 5 – it is that students' tendency to over-rely on linearity is extremely strong, deep-rooted, and difficult to break, even when substantial help is provided. This latter qualification applies not only to once-off forms of help (such as providing drawings, instructing to make drawings, or giving metacognitive hints), it even holds for more substantial help in the form of a 10-hour lesson series specifically dealing with this topic. The interview study (chapter 4) revealed that the phenomenon is due to a complex interaction of three factors: (1) the intuitiveness of linear relationships, (2) inadequate habits and beliefs towards mathematical word problem solving, and (3) content-specific elements.

The second explanatory factor – i.e., students' attitudes and their beliefs towards word problem solving – seems to be related to the kind of tasks that were used in the previously mentioned studies. In these studies, tests containing traditional, rather inauthentic, word problems were administered in a classical scholastic context. This even holds for the teaching experiment (chapter 5), wherein students' progress was assessed by tests comprising (only) school-like word problems, and even for follow-up study 3 (chapter 3), where the word problems were linked to fragments of the video on 'Gulliver's travels' shown before the test was taken. Although these word problems were embedded in a context made as authentic as possible, the effect on students' performances was disappointing. In this chapter, we report a last empirical study, in which we investigated the impact of a more radical operationalisation of authenticity, namely confronting students with an authentic, meaningful 'performance-based' task.

2. THEORETICAL BACKGROUND

In the mathematics education research community, it is repeatedly argued that tests containing traditional word problems may trigger in students a set of implicit rules and expectations established by the socio-mathematical norms of the classroom setting (see, e.g., Boaler, 1993; Cobb, Yackel, & McClain, 2000; Cooper & Harries, 2002, 2003; Lave, 1988, 1992; Palm, 2002; Reusser, 1988; Verschaffel, et al., 2000; Wyndhamn & Säljö, 1997). It may be suspected that students who over-relied on linearity in our previous studies did not invest sufficient mental effort and did not activate potentially helpful (meta)cognitive strategies in solving the problems because they assumed that they were dealing with routine word problems (see also the discussion section of chapter 4). They may, moreover, have excluded some considerations (e.g. that the obtained solution should look acceptable compared to the real-world situation that it refers to) and solution strategies (like checking the viability of a solution by making a sketch of the problem situation) – assuming that they were not required, or even not acceptable, in that context (Palm, 2002). Empirical evidence shows that for various kinds of modelling problems, students are more inclined to refrain from their stereotyped problem-solving behaviour and to include essential particularities of the real-world situation in their solutions when problems are disentangled from their scholastic chains and embedded in more meaningful, authentic tasks (e.g., DeFranco & Curcio, 1997; Lave, 1988; Nunes, Schliemann, & Carraher, 1993; Palm, 2002; Reusser & Stebler, 1997b; for an overview of such studies, see Verschaffel et al., 2000). The term 'authenticity' has no single meaning, but in general, authors use it to refer to tasks that are encountered in real, out-of-school, situations (or to truthful, high-fidelity simulations thereof in a school context), and they contrast it with 'inauthentic' mathematical tasks, which are the low-fidelity simulations typically encountered in school settings (Palm, 2002). Several authors (e.g., Cooper, 1992, 1994; Lave, 1988, 1992; Lesh, 1992; Palm, 2002; Verschaffel et al., 2000) mention essential differences between the two kinds of task. The main differences are summarised in Table 6-1 (for an elaborate discussion of the notion of authenticity related to mathematical tasks, see Palm, 2002). The more a task resembles the right column of Table 6-1, the more authentic it can be called.

Table 6-1. Characteristics of authenticity and inauthenticity in mathematical tasks

	Inauthentic	Authentic
Source of problem	Unlikely to occur in reality Imposed by external actor (teacher, textbook, task developer)	Occurring / likely to occur in reality Posed / encountered by problem solver
Problem formulation	Written Contextually limited, pre-formulated Contains all data and only required data, providing simplified data	Also oral and / or non-verbal Contextually rich, loosely defined Requires additional data, provides real(istic) data
Purpose and criteria	Providing the answer that teacher / task developer has in mind Task context is not helpful in clarifying criteria / adequacy of solution Additional, artificial requirements may be imposed	Find an adequate solution to a practical problem Task context gives additional clarification / specification of criteria All requirements originate from task context
Social/material conditions	To be solved alone Only scholastic tools (calculator, paper, pencil, ruler, …) Limited range of acceptable / plausible strategies Written response required	Other people can be involved Also concrete materials and other tools may be used Wider range of acceptable / plausible strategies Also oral / practical / non-verbal responses
Consequences	Correctness judged by others, approval or disapproval by teacher and other students	Feedback on degree of adequacy of solution by social / material environment

There are already studies that indicate a possible beneficial effect of increasing the authenticity of elements of a task on breaking students' tendency to inappropriately apply linear methods. Reusser and Stebler (1997b) used items from studies by Greer (1993) and Verschaffel, De Corte, and Lasure (1994), for which upper elementary students tend to give stereotyped, unrealistic answers. Among these, there were items for which students tended to overuse linearity, like the following:

- "John's best time to run 100 meters is 17 seconds. How long will it take him to run 1 kilometer?" (erroneous linear answer: $10 \times 17 = 170$ seconds)

- "A flask is being filled from a tap at a constant rate. If the depth of the water is 4 cm after 10 seconds, how deep will it be after 30 seconds?" (with a picture of a cone-shaped flask shown) (erroneous linear answer: 3 × 4 = 12 cm)
- "A man wants to have a rope long enough to stretch between two poles 12 meters apart, but he only has pieces of rope 1.5 meters long. How many of these pieces would he need to tie together to stretch between the poles?" (erroneous linear answer: 12 / 1.5 = 8 pieces)

Reusser and Stebler (1997b) gave these problems to students twice. First, as part of a typical scholastic paper-and-pencil test and, second, the day after, in a performance setting with concrete materials (e.g., pieces of rope, scissors and a meter stick for the rope problem or a cone-shaped flask and a jug of water for the flask problem) and with a clear performance instruction (namely, to investigate the problem using the materials, make a prediction about the answer, execute the task and write down their final answer). Students were remarkably less inclined to make the linear error in the performance setting. Compared to the paper-and-pencil condition, the percentage of correct answers increased from 12% to 47% for the runner problem, from 7% to 40% for the flask problem, and from 18% to 62% for the rope problem.

The goal of the present study was to investigate whether such an approach would also be effective in breaking students' linear reasoning tendency on problems about the effect of the enlargement of a figure on its area. Our previous investigations showed that the over-reliance on linearity in the domain of length-area problems is very general and extremely resistant to even strong kinds of help (see chapters 2 to 5), and we hoped to gain a deeper understanding of (the origins of) its resistance by investigating students' reactions in a different experimental context. Next to the mathematical domain of the problems, a second difference with the Reusser and Stebler (1997b) study is that for the problems in their study, students could rely on the linear model but needed to consider the outcome of their linear solution in the original problem context – which may indeed be facilitated by more authentic tasks (Palm, 2002). For the length-area problems in our study, however, the linear model is totally not valid, and a considerably more complex mathematical model needs to be applied instead. A third and final contrast with previous research on the authenticity effect is that we not only wanted to investigate the immediate effect of authenticity, but also whether the experience of correctly solving an authentic non-linear problem would have a beneficial effect on students' solutions on traditional school word problems on this topic as well.

Actually, Follow-up study 3 on the effects of authentic contexts and drawing activity (see chapter 3) had a comparable aim, namely to investigate the effect of task authenticity on students' over-reliance on linearity for problems on lengths, areas and volumes of geometrical figures. In that study, it was done by prefacing a paper-and-pencil test by an assembly of video fragments telling the story of Gulliver's visit to the isle of Lilliputians – where all lengths are 12 times smaller – and linking test items directly to these fragments. For instance, students in the 'authentic' condition got problems like "Gulliver's handkerchief has an area of 1 296 cm². What is the area of a Lilliputian handkerchief?", whereas control condition students received parallel versions formulated as traditional school problems not embedded in a contextually meaningful setting: "The side of square *Q* is 12 times as large as the side of square *R*. If the area of square *Q* is 1 296 cm², what's the area of square *R*?" Contrary to expectations, students in the 'authentic' condition did not outperform the control condition students. As we discussed in that chapter, watching a video and completing a paper-and-pencil test about that video was probably too weak an operationalisation of authenticity. With the present study, we wanted to test the effect of a more radical operationalisation of authenticity, in line with the characteristics mentioned in Table 6-1. Our major research questions were as follows. When students are confronted with a genuinely meaningful, performance-based mathematical task, do they approach it differently than a traditional school word problem, and does this possibly different approach also positively affect their performance on non-linear geometry problems? Also, if there is a positive effect of authenticity on students' task approach and performance, is there also a retention effect on their word-problem-solving behaviour and skills afterwards?

3. METHODOLOGY

3.1 Participants

The study was conducted with 93 6th-graders (five whole class groups from two medium-sized elementary schools, both located in Flanders and attracting average student populations). In 4th and 5th grade, these students had already encountered the concepts of area and volume and learnt how to calculate areas and volumes of various basic geometrical figures. In principle, the students possessed the necessary mathematical content knowledge and skills for solving all the non-linear area problems used in the study.

The investigation was conducted in three steps. First, a pretest was administered to all students. Next, students who made the envisaged linear error on the pretest were involved in an individual interview. Third, all participants got a post-test. Each step is explained in more detail below.

3.2 Pretest

A paper-and-pencil pretest with six word problems was given to students in their regular classroom by a member of the research team. Five of the word problems in the pretest were buffer items, intended to distract students from the actual focus of the study. One was the experimental item, aimed at detecting whether students tended to give a linear answer to problems about the effect of an enlargement on the area of a square.

Two versions of the test were constructed. Half of the students received version A and half received version B. In the post-test (see below), each student got the other version. The experimental items in test version A and B were equivalent in the sense that both dealt with a square of which the sides were doubled:

- Version A: "John needs 15 minutes to paint a square ceiling with a side of 3 meters. Approximately how much time will he need to paint a square ceiling with a side of 6 meters?"
- Version B: "Carl needs 8 hours to manure a square field with a side of 200 meters. Approximately how much time will he need to manure a square field with a side of 400 meters?"

3.3 Interview procedure

Two or three days after the pretest, the students who made a linear error on the experimental pretest item were individually taken out of the classroom for an in-depth interview. These students were asked to solve one non-linear problem, this time about the effect of tripling the lengths of the sides of a square on its area. The problem was offered in one of three different ways, depending on the experimental condition that the student was assigned to. Assigning students to conditions was done by matching, based on students' mathematical performances on formative school assessments during the previous months.

Students in the *S-condition* ('Scholastic' condition) received a sheet with the problem formulated as a traditional word problem, presented in a similar way as the non-linear word problem in the pretest and post-test:

"Recently, I made a doll's house for my sister. One of the rooms had a square floor with sides of 12 cm. I needed 4 square tiles to cover it. Another floor of the doll's house was also a square, but with sides of 36 cm. How many of those square tiles did I need to cover it?"

The *D-condition* ('Drawing' condition) was exactly the same as the S-condition, but this time the sheet also contained a drawing to scale of the small and large figure, as shown in Figure 6-1.

12 cm 36 cm

Figure 6-1. Drawing offered in the D-condition

Both in the S- and D-condition, the answer that the student wrote down was registered as their final answer.

In the *P-condition* ('Performance task' condition), the problem was presented as a 'performance task'. Students were involved in the actual problem situation with real materials (the small dollhouse floor, 4 tiles and the large doll's house floor) and were asked to perform an authentic action. The task was introduced as follows:

> "I have a little sister, and currently I am making a doll's house for her. Here, you can see the floor of one of the rooms. Can you tell me its shape? [The pupil tells that it is a square.] Let's measure it. [The pupil observes that the sides are 12 cm long.] I have some tiles that we can use to cover that floor. Can you do that? [Pupil puts 4 tiles on the small floor.] Indeed, we need 4 tiles to cover that floor.
>
> I also brought another floor of the doll's house. As you can see, it is also square. Let's measure it as well. [Pupil observes that the sides are 36 cm.]. In a few moments, we are going to put tiles on this large floor as well. Now, think about how many tiles we will need to do that. If you have decided you can go to the table over there and fetch exactly enough tiles."

In the P-condition, the number of tiles that the students actually fetched was registered as the final answer. At the end of the interview, P-condition students were allowed to put the tiles that they fetched effectively on the large floor (but note that their final answer was already registered before this happened). Unavoidably, this revealed whether their answer was correct or not. When students observed that they did not bring enough tiles, they could fetch additional tiles to cover the entire floor.

It is clear that the task settings in the S- and D-condition have strong resemblance to the inauthentic task characteristics (left column of Table 6-1), whereas the P-condition corresponds more to an authentic task setting. Remark that for our research goal, strictly spoken the experimental design would only need to include a S-condition and a P-condition. Nevertheless, a D-condition was included for the following reason. Students in the P-condition not only received the non-linear problem as a performance task instead of a school-like word problem, but also the problem presentation in the P-condition gave additional visual support while this was not present in the S-condition. By including a D-condition – which provided equivalent visual support as the P-condition but which kept the scholastic setting from the S-condition – we could appropriately control for this factor.

At the beginning of the interview, students were told that they could solve the problem in whatever way they wanted and use all materials available (pencil, paper, ruler, pocket calculator, in the D-condition also the drawings, and in the P-condition the small and large floor and the four tiles that were already available to them in this phase). Students were asked to think aloud while solving the problem. When they immediately came up with an answer without verbalizing their thinking, the interviewer asked some standardized probing questions ("Where does that answer come from?", "Can you explain how you found that answer?"). These questions were posed only after the final answer was registered, to ensure that these probing questions would not affect students' answers.

At the end of the interview, students were also asked to indicate on a five-point scale how certain they were about the correctness of their answer ('certainly wrong', 'probably wrong', 'no idea', 'probably correct' and 'certainly correct') and to justify this. Evidently, students in the P-condition had to do this before they were allowed to check their answer by putting the tiles on the large floor. Interviews were videotaped and transcribed for later analysis.

3.4 Post-test

Either one or two days after their interview, the students solved a post-test. This test was given to all 93 students who solved the pretest rather than only to the interviewed students, to avoid students becoming too suspicious about why they were (or were not) tested. Only the results of the students who were involved in the actual study (i.e., the ones who made a linear error on the pretest and who were individually interviewed) are reported here. The post-test again was taken in students' regular classroom under supervision of the researcher. As mentioned before, students who got test version A at the pretest got version B at the post-test and vice versa.

4. RESULTS

4.1 Pretest and individual interviews

Altogether, 72 of the 93 students gave a linear answer to the experimental item at the pretest. They were all distributed over the three interview conditions (N = 24 in each condition). Table 6-2 provides the answers (categorized as correct, linear error or other error) and solution times (i.e. the time needed to find an answer after the problem was introduced) in the different interview conditions. As shown in the table, there was a statistically significant impact of the interview condition on students' answers[35].

Table 6-2. Overview of answers and average solution times (in seconds) in each interview condition

Condition	Answer	Frequency	Average solution time
S-condition	Correct	1	120
(N = 24)	Linear	21	50
	Other error	2	115
D-condition	Correct	16	139
(N = 24)	Linear	8	61
	Other error	0	/
P-condition	Correct	20	76
(N = 24)	Linear	2	29
	Other error	2	131

In the *S-condition*, nearly all students (21 of 24) committed the linear error again on the interview item. This confirms students' deep-rooted tendency to apply linear methods when solving problems about areas of enlarged figures as observed in all previous studies. Two students committed another error and one student found the correct solution (in contrast with the pretest, this student now made a drawing himself, which apparently helped him to find the correct solution). In the *D-condition*, performances were

[35] Fisher exact test: $p < .00015$

considerably better[36]. Here, 16 of 24 students found the correct answer during the interview. In fact, we had expected that visual support as such would not be very helpful for most students, since studies reported in chapters 2 and 3 showed that providing drawings hardly affected performance (because students often neglected them). Here, students actually used the drawing quite often and their solution process clearly benefited from it. Possibly, students in the current study felt more obliged to do so because they were involved in an individual interview, whereas the studies reported in chapters 2 and 3 used collective, written tests, which may have elicited a different 'experimental contract' (Greer, 1997, p. 305). Despite the beneficial impact of drawings for 16 students, it should be noted that 8 of 24 D-condition students still gave a linear answer. Compared to offering drawings, transforming the non-linear problem into an authentic performance task was more beneficial for students' solutions. In the *P-condition*, 20 students gave the correct answer, and only 2 students reasoned linearly (2 students made another error). A contrast analysis of the D- and P-condition answers showed that also this difference was statistically significant[37].

In sum, almost all students in the S-condition continued to make the linear error they had made during the pretest. Providing a drawing had a positive effect, but still one third of the D-condition students made the linear error. Offering the problem as an authentic performance-based task was even more beneficial. In the P-condition, linear errors were nearly absent.

Given that not only the P-condition but also the D-condition yielded a relatively large number of correct solutions (20 and 16 out of 24 students, respectively, versus only 1 out of 24 in the S-condition), we also performed a more fine-grained comparative analysis of these correct solutions in these two groups, with a view to determining possible differences in the way the correct solutions had originated and in students' feelings of certainty of the correctness of the answer. Before answering these questions, we emphasize that only a few students were able to clearly verbalise their problem-solving processes. Therefore, this comparison is necessarily based mainly on the other data we were able to collect and analyse about students' problem-solving process, namely response times (cf. Table 6-2), observable use of the drawings or materials, and certainty scores (see the overview in Table 6-3) with accompanying justifications. But before presenting these contrastive data, we illustrate, in Table 6-4, the major differences between the correct solution processes in the two conditions by means of two typical interview protocols.

[36] Fisher exact test: $p < .00015$ (for the contrast between the S- and D-condition)

[37] Fisher exact test: $p = 0.0412$

Table 6-3. Distribution of 'certainty' scores given by students answering correctly in the D- and P-conditions

Condition		Certainly wrong	Probably wrong	No idea	Probably correct	Certainly correct
D-condition	Frequency	0	0	3	10	3
($N = 16$)	Percent	0	0	19	62	19
P-condition	Frequency	0	0	2	10	8
($N = 20$)	Percent	0	0	10	50	40

Table 6-4. Illustrative thinking aloud protocols of students in the D- and P-conditions (both finding the correct answer)

D-condition

Deborah (solution time: 194 seconds)

Deborah:

[Silence of about 60 seconds. Rereads the word problem several times.] "So … I think I should … Is this the drawing of the small and large floor? [Measures sides of small and large square.] The small floor is 12 cm and the large 36. So that's 3 times … eh … 3 times 4 tiles is 12. No, it's 3 times here and 3 times here, that's 9 times. You need 9 tiles … Wait, let me read it again. [Reads the word problem and thinks for a long time.] I don't know how I should calculate it. Maybe here … there's 4 tiles and … yes! [Draws 9 of the small floors in the large floor.] 9 times more, so I do 9 × 4… 36"

Interviewer:

"So, 36 is your answer. Can you tell me how certain you are that that is the correct answer?"

Deborah:

[Chooses 'Probably correct'] "I'm never sure about myself, and here, I don't trust it. Maybe there's something wrong with the drawing, you know. It could be a tricky question."

P-condition

Marlies (solution time: 34 seconds)

Marlies:

[Immediately takes the small floor and fits it several times on the large floor.] "It's 9 times this small one, which has 4 tiles, so 36 tiles." [Leaves immediately to fetch 36 tiles.]

Interviewer:

"So you brought 36 tiles. Can you tell me how certain you are that that will be the correct amount of tiles?"

Marlies:

[Chooses 'Certainly correct'.] "It is just correct, I showed you that it's 9 times more. Why would I need to doubt about it? I am just sure."

In the D-condition, all 16 students who found the correct answer needed relatively long times (on average 139 seconds, cf. Table 6-2), and had to rely extensively on the drawing. All of them used the drawing in some way (mostly by drawing the tiles on the large floor drawing as well, or by dividing the large floor in small ones), but this idea often came up rather late in the problem-solving process, after a long time of thinking and (re)reading the problem. Five students initially applied a linear method, and abandoned it only after they felt that they needed to "do something with the drawing too". Once given the correct answer, many students were still not totally convinced about its correctness (cf. Table 6-3). When referring to sources of *certainty*, students from the D-group (even those indicating 'certainly correct') never expressed the idea that the answer was *necessarily* correct. Justifications rather related to (the strength of) their own mathematical abilities, to a feeling of "having conducted the required calculations", and/or to the fact that a recalculation had confirmed the correctness of the first calculation. When referring to sources of *uncertainty*, students indicated that various reading, interpretation or calculation errors might have occurred, or that they might have overlooked a critical aspect of the problem situation.

The above observations from the D-condition solution are in contrast with the processes observed in the P-condition. In that condition, students needed much less time to respond correctly – on average only 76 seconds (cf. Table 6-2). In most cases, they immediately and spontaneously started manipulating the materials, working towards the correct solution (figuring out rather quickly that 6×6 tiles fit on the large floor, or that the small floor fits 3×3 times on it). Also, none of them initially applied a linear method. Three students from the P-condition gave the correct answer immediately once the problem situation was explained to them (while this never happened in the D-condition). These students just 'saw' the correct solution at a glance. Once they had given the answer, students in the P-condition were also more convinced of its correctness than the students from the D-condition (cf. Table 6-3). When justifying their certainty rating, they did not refer to just having done the required calculations or to their own mathematical abilities (like many students in the D-condition did). Rather, they used expressions stating that the answer was *necessarily* correct, arguing that the solution was logical or evident whereas any alternative was just unthinkable, or claiming that they could just *see* the correctness of the answer.

4.2 Post-test

Table 6-5 shows the answers on the non-linear post-test item for the different interview conditions and interview responses separately. The post-test aimed at determining whether students who had solved the non-linear item correctly during the interview, would do this on the post-test (taken one or two days later) as well. Interestingly, this occurred only rarely. The single S-condition student who found the correct solution solved the post-test problem correctly too, again by making a drawing. But in the other two conditions hardly any effect of the interview experience was found. While 16 of the 24 D-condition students profited from the drawings to find the correct answer in the individual interview, all of them again reasoned linearly on the post-test. And whereas 20 of the 24 students found the correct answer in the P-condition interview, only two of these 20 students solved the post-test item correctly (and all others, except one, again made a linear error).

Table 6-5. Overview of students' answers at the post-test item in relation to their answer during the interview

Condition	Answer during interview	Total	Answer at post-test		
			Correct	Linear	Other error
S-condition	Correct	1	1	0	0
(*N* = 24)	Linear	21	1	20	0
	Other error	2	0	2	0
D-condition	Correct	16	0	16	0
(*N* = 24)	Linear	8	1	6	1
	Other error	0	/	/	/
P-condition	Correct	20	2	17	1
(*N* = 24)	Linear	2	1	1	0
	Other error	2	0	0	2

Finally, of the four P-group students who failed to solve the problem by themselves during the interview (appearing in the two bottom lines of Table 6-5) but who were allowed to put their tiles on the floor and see the correct answer at the end of the interview, only one student solved the non-linear post-test item correctly. The other three gave the same answer on the post-test as during the interview.

5. CONCLUSIONS AND DISCUSSION

In all previous studies about students' over-reliance on linearity to solve problems about the area and/or volume of enlarged figures (chapters 2 to 5), we observed this phenomenon by means of tests containing traditional, scholastic word problems. The current study revealed that the choice of this method has an important impact on their solution behaviour. When students who made a linear error on a scholastic pretest were involved in an interview with a more authentic, performance-based task, they were considerably less tended to over-generalise the linear model to that task, compared to students who got a traditional word problem or even a word problem with drawings. Based on the interview study reported in chapter 4, the phenomenon of improper linear reasoning was explained by different aspects of students' knowledge base. The present study highlights the important role of the scholastic context as an explaining factor for students' overuse of linearity. It clearly support Boaler's (2000, p. 118) argument for a consideration of the 'macro context' in which research is conducted, for

> extending our focus beyond the concepts and procedures that students learn to the practices in which they engage as they are learning them and the mediation of cognitive forms by the environments in which they are produced.

When students solve traditional word problems, we can indeed observe a clear and deep-rooted tendency to over-rely on linearity, but such a methodology arguably reveals more about the specific kind of 'practices' (Boaler, 2000, Lave, 1988) that students engage in when they are in a traditional school word problem-solving context than it tells about their actual capacities for mathematical reasoning:

> The activity of solving word problems and the contents of word problems in school are not the same as 'the same' activity or contents embedded in other systems of activity in other parts of life (Lave, 1992, p. 89).

In general terms, our results endorse the conclusion of other studies that increasing the authenticity of a task positively affects students' responses (e.g., Cooper & Harries, 2002, 2003; DeFranco & Curcio, 1997; Lave, 1988; Nunes et al., 1993; Palm, 2002; Reusser & Stebler, 1997b; Verschaffel et al., 2000). More specifically, our study showed – as was already suggested by Reusser and Stebler's (1997b) study described above – that students' tendency to overuse linear methods declines significantly when problems are disentangled from their scholastic chains and are embedded in a meaningful task setting. But our study went further than just providing additional empirical evidence for this general claim, in two important ways.

First, Reusser and Stebler (1997b) focused on problems for which the linear model provides a more or less adequate approximation. For the runner item, for example, students should be aware that their linear calculations ("it takes 10 × 17 seconds to run 10 × 100 meters") are correct, but only can serve as a first approximation while the real answer would be somewhat higher because runners become tired. Or, for the ropes item, the students should realize that linear calculations were indeed relevant (8 pieces of rope each being 1.5 meters long indeed have a total length of 12 meters) but the knots when connecting the pieces of rope would probably cause the need for one or maybe two additional pieces. With the present study, we showed that a meaningful, authentic task setting can also enable students to avoid the use of the linear model in a situation where it is simply not valid (even not as a tentative approximation), and to select or develop a considerably more complex mathematical model instead.

Second, our findings point out that the effect of presenting the problem as a meaningful, performance-based task was only a temporary and context-specific one with only marginal impact on students' future problem-solving behaviour. In a post-test, which again contained traditional word problems and which was taken only a few days later, nearly all students reverted to the linear answer. The explanation for this remarkable observation is still an open question. Maybe the relapse was due to the fact that students were unable to see the structural mathematical connection between the problem situation offered in the interview, on the one hand, and the problem situation described in the experimental item, on the other hand (Reed, 1999). But, alternatively, students may just have neglected searching for such structural mathematical connection because they did not consider that kind of action as relevant in the post-test setting. Unconsciously or even deliberately, they may have taken into account the different experimental setting of the individual interviews and the collective word problem test, and therefore behaved according to a different 'experimental contract' (Greer, 1997). This issue needs to be clarified by future research, contrasting students' reactions to post-test tasks both in a paper-and-pencil test and in an individual interview setting. Retrospective in-depth interviews specifically dealing with students' perceptions while being tested or interviewed may also bear upon our better understanding of how the different experimental settings influence students' responses.

Although this issue requires further research, our findings already suggest – as Boaler (1993, p. 23) phrased it – that

> assumptions regarding enhanced understanding and transfer as a result of learning in context may be oversimplistic.

The use of authentic mathematical tasks may strengthen students' motivation and interest and make them approach problems more meaningfully and thoughtfully, but this does not imply that students will be inclined and able to establish connections between different problem situations with an equivalent mathematical structure. Instructional interventions may find a valuable starting point in performance-based tasks like the one developed in the current study, but explicit instructional efforts seem anyhow necessary to bridge the gap between students' informal, context-bound work and the intended generalisable mathematical insights that they can rely upon on later occasions.

Chapter 7

PSYCHOLOGICAL AND EDUCATIONAL ANALYSIS

1. INTRODUCTION

This final chapter consists of two main sections. In the first section, we analyse and discuss the psychological and educational factors that seem to lie at the roots of the occurrence and persistence of the over-reliance on linearity. This analysis is based on the overview of cases of unwarranted applications of linear properties and representations in diverse mathematical domains that we discussed in chapter 1, and on the conclusions of the series of studies in the domain of geometry reported in the central chapters of this monograph. We will argue that explanatory elements are found in (1) the strongly intuitive and heuristic nature of the linear model and its omnipresence in our everyday life, (2) students' experiences in the mathematics classroom having an impact on the further development and the persistence of this 'natural' tendency, and (3) elements or 'triggers' related to the specific mathematical problem situations in which linearity errors occur.

In a second section, we formulate recommendations for the improvement of educational practice. First, we suggest some concrete modifications of current practice, specifically related to the teaching of linearity and the concepts of area and volume. These modifications would constitute an important step forward in breaking students' tendency towards improper linear reasoning. Second, we formulate a more comprehensive response to the research findings reported in this monograph, wherein the opposition against the linear tendency is embedded into a more general so-called 'modelling perspective towards mathematics education'.

2. SEARCHING FOR THE ROOTS OF THE OVERUSE OF LINEARITY

The aim of the present section is to create order in our knowledge base on students' overuse of linearity, and to identify, unravel, and compare the different explaining factors. This overview will not result in one overarching theoretical framework that incorporates and integrates the various elements that are at the roots of the over-reliance on linearity. Taking into account the current stage of the research with respect to this phenomenon, it seems premature to attempt formulating such an integrated framework. But we think that bringing together the existing evidence and interpreting it from different angles can already be a valuable step towards that goal. We have grouped the factors explaining the over-reliance on linearity into three major categories. The first type of explanation can be found in the ubiquitous applications of linearity in everyday life and in the fact that human cognition tends to be biased and intuitive, as has been claimed by several psychological theories. A second category of explanatory factors lies within the practices of mathematics education at school and in the experiences that students have within the school system. These factors relate to the scholastic system reinforcing students' 'natural' inclination towards linearity and contributing to its further development. We will analyse these factors from a socio-constructivist perspective on mathematics teaching and learning and the situated cognition framework. And third, a scrutinized analysis of the mathematical (sub)domains wherein specific errors occur and of the mathematical concepts that are related to the committed errors may provide additional insight into the reasons why students over-rely on linearity in particular cases. Before elaborating more deeply on each of these three elements, we want to stress that most often manifestations of unwarranted proportional reasoning result from a complex interplay of several explanatory factors.

2.1 Linearity, intuitiveness and everyday life situations

The first and most important explanation for the tendency to over-rely on linearity seems to lie in the ubiquitous utility of linear relations in our everyday life and the observation that human cognition often tends to rely on intuitive instead of analytical modes of thinking, as argued for example by Fischbein in his theory on intuitions and schemata in mathematical reasoning (Fischbein, 1987, 1999), or by cognitive psychologists like Evans (2003), Sloman (1996) and Stanovich and West (2000) in their dual process theory.

It has often been argued that when the linearity concept is being developed, it tends to receive a more and more self-evident, even intuitive, status. Rouche claimed that

> C'est l'idée de proportionnalité qui vient d'abord à l'esprit, parce qu'il n'y a sans doute pas de fonctions plus simples que les linéaires[38] (1989, p. 17).

Indeed, from a mathematical point of view, a function of the form $f(x) = ax$ is one of the most simple, elegant relationships that can occur between two variables. Already at a very young age children begin to master some elementary proportional reasoning skills in certain types of problem situations. Situations in which a relatively small integer proportionality factor is salient can start to elicit quite naturally a multiplicative 'k times a–k times b' reasoning or, for example, an '$a + a$, so $b + b$' building-up reasoning (which both rely on the linear characteristics of the problem situation). But as early as certain linear ideas appear, the first cases of their overuse can show up as well. Research-based evidence can be found, for example in the observation that about half of the students in Grades 2 and 3 – i.e., well before systematic instruction into the topic of proportionality – gave a proportional answer to the word problem

> It takes 15 minutes to dry 1 shirt outside on a clothesline. How long will it take to dry 3 shirts? (Van Dooren et al., 2005).

When looking at several cases of the overuse of linearity, the connection with Fischbein's (1987, 1999) theory of intuition in mathematical reasoning is striking. Fischbein described intuitive knowledge as a type of immediate, implicit, self-evident cognition, based on salient problem characteristics, leading in a coercive manner to generalisations, generating great confidence and often persisting despite formal learning. Intuitions are resistant to change because they are profoundly related to our adaptive system, which is shaped by early and repeated experiences. Fischbein (1999, p. 45) indeed considered proportionality as a schema that is part of our adaptive system, and he commented that

> The schema of proportionality is, in the Piagetian terminology, an operational schema, that is, a very general, influential one. It develops with age and in its full, quantitative form, it manifests itself during the formal operational stage.

Numerous everyday life situations indeed involve linearity (Akatufba & Wallace, 1999; Spinillo & Bryant, 1999), and even several of our most primitive experiences like counting objects are reflections of linearity. In this respect, Rouche (2001) made a strong point in arguing that

[38] The idea of linearity immediately comes to the mind, because there are no simpler relations than linear ones.

> Si presque toutes les mathématiques sont linéaires, n'est-ce pas parce que le linéaire est la structure la plus conforme à l'esprit de l'homme ? (l'esprit donné congénitalement ou modelé par les pratiques les plus familières)[39].

Therefore, it does not come as a surprise that several authors refer to the intuitiveness, simplicity, or elegancy of linear ideas. For instance, Stacey (1989, p. 162, our emphases) concluded on her observation of students' unwarranted use of the $f(ka) = k f(a)$ and $f(a + b) = f(a) + f(b)$ properties of linear functions (see chapter 1) that

> the models associated with direct proportion suggest themselves to students *for strong cognitive reasons*. When such an idea is found, students may be reluctant to question it, both because of its effectiveness in supplying answers (and any answer is better than none!) *and because of its simplicity*.

Clear empirical indications for the intuitive application of linearity are found in the in-depth investigation reported in chapter 4. Students opted very quickly – almost immediately after confrontation with the problem – for a proportional solution method. When questioned afterwards, they could hardly explain why that method should be used for that particular problem. At the same time, they showed great confidence, without feeling any need to justify it. Although the students had all the required mathematical content knowledge to respond correctly, they were remarkably persistent in their erroneous proportional answer, even after being confronted with strong contradictory evidence. So, the characteristics of intuitive cognitions described by Fischbein (1987) clearly apply here.

Besides Fischbein, the distinction between intuitive and analytical modes of thinking has been expressed and studied by many cognitive psychologists, especially by advocates of dual process theories of thinking (e.g., Evans, 2003; Sloman, 1996; Stanovich & West, 2000). These researchers posit the existence of two different reasoning systems. The first system (often called S1 or the heuristic system) is characterised as automatic, associative, unconscious, and requiring less effort and cognitive resources. This system produces quick reactions and is often based on salient features of the problem situation and on stored 'prototype' situations (Sloman, 1996). The second system (S2 or the analytical system) is characterised as controlled, deliberate, and requiring much resources from our working memory. When

[39] If almost all mathematics are linear, isn't it because linearity is the structure that conforms most to the human mind? (the mind that is innate or modeled by the most familiar practices).

people fail to provide the normatively correct answer to reasoning tasks (e.g., answering "10 cents" to the problem "A baseball bat and ball cost together one dollar and 10 cents. The bat costs one dollar more than the ball. How much does the ball cost?", see Kahneman, 2002, p. 451), this is attributed to the pervasiveness of S1 whereby S2 fails to intervene in its role as a critic. It is argued that S1 provides so many useful responses in various situations that people tend to develop biases in the sense that they also give S1 responses to reasoning tasks that require analytic (S2) processing. An important implication of this theory is that erroneous answers do not necessarily need to be interpreted in terms of shortcomings in the relevant (mathematical) knowledge base (S2) – in which case one would speak of misconceptions – but often simply are the consequence of S1 activity before S2 could act. The phenomenon of the overuse of linearity seems to fit remarkably well to these dual process theories. When students experience the usefulness of linearity in various settings, a kind of 'linearity heuristic' may become part of S1 and start to act strongly on students' reasoning, as it produces quick and often correct responses. In many cases, students' overuse of linearity is not a consequence of deficiencies in their mathematical knowledge (S2), but rather the result of a tendency to react too quickly and too impulsively (S1).

Finally, there also seems to be a parallel between students' illicit application of proportionality and the 'intuitive rules' described and studied by Tirosh and Stavy (1999a, 1999b). Building on the work of Fischbein (1987), these authors claim that there are some common, intuitive rules that come into play when students solve problems in mathematics and sciences. Two such rules are manifested in comparison tasks: 'more *a*–more *b*' (when in a problem situation, a salient magnitude *a* increases, students reason that another magnitude *b* increases as well, e.g. "a rectangle with a larger perimeter has a larger area") and 'same *a*–same *b*' (when in a problem situation, a salient magnitude *a* remains the same, students reason that another magnitude *b* does not vary either, e.g., "triangles with the same perimeter have the same area"). Several examples that Tirosh and Stavy mention on the application of the 'same *a*–same *b*' rule are in fact cases of improper proportional reasoning (e.g. the belief that it is equally likely to get 2 or more heads in 3 coin tosses as to get at least 200 heads in 300 tosses, see also Van Dooren et al., 2003). The link between the overuse of linearity and the intuitive rules theory also became clear in the in-depth investigation (reported in chapter 4). Several students argued that a *larger* figure has *more* area, and because during the enlargement the figure keeps the *same* shape, everything enlarges with the *same* factor, leading to a '*k* times *a*–*k* times *b*' reasoning. Nevertheless, the relation between the over-reliance on proportionality and the intuitive rules theory developed by Tirosh and Stavy needs to be examined more thoroughly. A recent empirical study on intuitive rules with 10th- and 12th-graders (Van Dooren, De Bock, Weyers, & Verschaffel, 2004) already pointed out that students who answered in line with the 'same *a*–same *b*' intuitive rule for this type of problem often explained their answers in terms of direct proportionality (and not always explicitly referred to a 'same *a*–same *b*' rule), while the 'more *a*–more *b*' rule proved to be only very rarely influential in these older students' reasoning processes.

2.2 The effects of schooling

Although the previous section suggests that there is a 'natural' tendency to reason proportionally, there is ample evidence that the practices of mathematics education at school and the experiences that students have within the school system have an impact on its further development and contribute to its persistence. In a socio-constructivist view on mathematics teaching and learning (see, e.g., Cobb, 1994), students are not passive receivers of ready-made mathematics. Rather, they are active participants in the teaching-learning process who construct mathematical tools and insights by making sense of their experiences and practices, in interaction with others and with their environment. As learners encounter new situations, they look for similarities and differences with their own cognitive schemata, and draw their own conclusions based on this interaction. This learning process also has a social dimension, as the interactions between teachers and students during the mathematics lessons – and, we would argue, also the processes of students being individually engaged in mathematical tasks presented within a scholastic context, including tests – are based on a complex set of (mainly implicit) rules and mutual expectations, often denoted by the term 'didactical contract' (Brousseau, 1997) or the closely related concept of 'sociomathematical norms' (Cobb, Yackel, & Wood, 1992) (see also Verschaffel et al., 2000).

Such a perspective on mathematics teaching and learning brings up numerous explanations for students' tendency to over-rely on linear methods. A first explanatory element is that at certain moments in the mathematics curriculum extensive and almost exclusive attention is paid to linearity or to one of its properties or representations. Often, there is a strong focus on the technically correct and fluent execution of certain procedures related to these properties and representations, without explicitly and systematically questioning their applicability. The consequences are for example shown in the study by Van Dooren et al. (2005) in the field of arithmetic, where a considerable increase was found in the number of illicit proportional answers from 3rd grade to 5th grade, i.e. the period during which there is extensive attention to the acquisition of the proportionality scheme (see chapter 1). For some problems, it was found that 3rd-graders even gave more correct answers than 5th- or 6th-graders, due to the increased tendency in these latter groups to use proportional models 'anywhere'. These data clearly suggest that, with respect to proportional reasoning, these students had acquired 'routine expertise' instead of 'adaptive expertise' (Hatano, 1988).

A particularly salient case relates to the observation that, in the regular mathematics curriculum, proportional reasoning is mainly practised by means of so-called missing-value problems. In this problem type, three numbers (a, b, and c) are given, and the problem solver is asked to determine an unknown number x. In a proportional problem, this unknown x is the solution of an equation of the form $a/b = c/x$. The majority of proportional reasoning tasks students encounter in the upper grades of the elementary school and the lower grades of secondary school is formulated in a missing-value format (Cramer et al., 1993) – many textbooks even exclusively rely on this format – and a lot of attention is paid to the use of arrow scheme representations as a powerful and efficient tool to tackle this kind of problem. At the same time, non-proportional problems stated in a missing-value format are very rare. When solving large numbers of missing-value problems exclusively with proportional solution methods, while a justification for choosing that method is generally not required, this experience will be reflected in the mathematical conceptions and activities that students develop. The second follow-up study in chapter 3 on the effects on problem formulation empirically confirmed that the missing-value format is an important explanatory factor for the occurrence and persistence of the over-reliance on linearity, as rephrasing the usual missing-value problems into comparison problems proved to be a substantial help for many students to overcome their tendency towards improper linear reasoning. Nevertheless, more than half of the rephrased non-proportional problems were still solved erroneously which indicates that the missing-value format can only partly explain the phenomenon under consideration.

Students' classroom experiences also have a more general influence on their (mathematical) problem-solving habits and beliefs, and these habits and beliefs have in their turn an undisputable role in the occurrence of the over-reliance on linearity. Several investigations (for an overview see Verschaffel et al., 2000) have shown that students often perceive mathematical problem solving as a puzzle-like activity with little or no relation to the real world. They tend to use superficial cues to decide which operation is required in a certain problem context (e.g., Ben-Zeev & Star, 2001; Hinsley, Hayes, & Simon, 1977; Schoenfeld, 1988). Also in several empirical studies reported in this monograph (see, e.g. chapters 2, 3 and 4), it was found that the over-reliance on linearity was due to students' reluctance to use some powerful heuristics (e.g., they rarely made or used a sketch or drawing to tackle a non-trivial problem) and to shortcomings in their metacognitive knowledge and skills (leading to a low self-monitoring while problem solving, being quickly satisfied with a solution, etc.). The fact that students in these studies exhibited such superficial behaviour may come as no surprise, as the non-linear tasks that were used in these studies were generally offered as traditional, school-like word problems. The impoverished and stereotyped diet of standard word problems, that the students had encountered day-by-day in their mathematics lessons and tests in the past, certainly shaped their problem-solving tactics and their accompanying beliefs about the intent, structure, and assumptions of solving mathematical word problems (Verschaffel et al., 2000). Or as Lave (1992, p. 89) argued, students' behaviour when solving word problems can be considered as a distinctive 'practice' with its own norms and habits:

> the activity of solving word problems in school is not the same as 'the same' activity or contents embedded in other systems of activity in other parts of life.

Clear-cut evidence for the role that the traditional word problem format plays in the occurrence of students' overuse of linearity was provided in chapter 6. When students were offered the same task not as a traditional word problem but as an authentic performance-based task, students worked immediately and promptly towards a correct solution for a non-linear problem, and linear reasoning disappeared almost entirely. But when, a few days later, the same students solved non-linear problems that were very similar to the performance-based task but that were again presented in a traditional format, nearly all students reverted to the erroneous linear solution.

2.3 Content-specific effects

The two previous sections dealt with two explanatory elements of a more general nature, in the sense that both the first factor (the role of intuitive, non-analytic reasoning) and the second one (the impact of schooling experiences) can account for a diversity of manifestations of illicit proportional reasoning in a variety of content domains. We now will argue that, in many cases, content-specific elements also contribute to students' overuse proportionality – which may not be surprising because (1) students have many difficulties with particular basic concepts and, more specifically, (2) certain concepts themselves are strongly related to proportionality. It is impossible to exhaustively describe all content-specific issues that may cause or facilitate the overuse of linearity, so we will illustrate the role of such effects for the context central to this monograph: the overuse of linearity related to the effect of a linear enlargement/reduction of a geometrical figure on its area and volume. Besides the observation that many students have difficulties with the notions of area and volume (Outhred & Mitchelmore, 2000; Rogalski,

1982; Tierney et al., 1990) – which may explain why so many errors occur on area or volume problems – there is also a content-related reason why students may erroneously fall back on *linear* methods. In order to maintain the same shape, an enlarged or reduced figure necessarily is enlarged or reduced in a *linear* way, i.e., the figure becomes k times higher and moreover also k times wider (and spatial figures k times deeper). In other words, when students approach a problem about an enlarged or reduced figure, their thinking is unavoidably pointed towards the linear changes in the described situation. For example, a student in the interview study reported in chapter 4 explicitly argued: "The height *and* the width of the figure are tripled, so the area is tripled too." His statement suggests that he clearly had an adequate mental representation of the problem situation and was not simply focussing on superficial characteristics of the problem formulation (that only mentioned the height of the figure). It was precisely this (appropriate) mental representation that convinced him of the correctness of his linear solution.

Also in the domain of probability, content-specific issues stimulate the occurrence of the overuse of linearity. In the section on probabilistic reasoning reported in chapter 1, we already stated that numerous misconceptions occur in that domain. More importantly, we also argued there that the concepts of 'chance' and 'proportion' are very strongly related (Fischbein & Gazit, 1984; Lamprianou & Lamprianou, 2002; Truran, 1994). For example, in situations with equally probable outcomes, the probability of a success can be estimated by means of the ratio of the number of successful outcomes and the number of trials, and proportional reasoning is required to know that the chance for drawing a white ball is the same if one urn contains 20 white and 40 black balls and another urn contains 100 white and 200 black balls. Finally, as probabilistic situations always deal with intrinsic uncertainty, immediate and conclusive feedback on the correctness of one's assumptions is often unavailable. It can be easily and convincingly illustrated that the area of a circle is not doubled if its diameter is doubled, but it is much more difficult to show that the probability of getting at least a six is not doubled if the number of rolls with a die is doubled.

Similar analyses can be made for other situations where student over-rely on linearity, although often there is not enough research-based information about how students understand, or misunderstand, the situations wherein they apply linearity.

2.4 Perspectives for further research

Although this section gave an extensive overview of the possible roots of students' tendency to over-rely on linearity, further research is needed to deepen our understanding of the phenomenon. In each of the groups of explanatory elements that were discussed above, there seem to be many perspectives for further research.

First, the suspicion (arising from the dual process theory) that linearity can be conceived as an automatic, pervasive heuristic system based on prototypes (S1) needs to be empirically verified. It should also be investigated how and why the analytic, effortful processing system (S2) so often fails to intervene in its role of monitor and critic in mathematical problem-solving situations. Typically, students in these situations are expected to think hard and for a longer period, and be able to justify their answer (see, e.g., Pólya, 1945; Schoenfeld, 1992), whereas the tasks examined by dual process theorists are generally more related to rather simple everyday life situations. The claim that linearity – according to this theoretical framework – underlies a first and quick response easily leads to the hypothesis that situations requiring a more considered response might elicit less unwarranted proportional answers.

Second, we need to know more about students' intuitive understanding of non-linearity and how this understanding develops with age and schooling. Recently, Ebersbach and Resing (2006) provided scientific evidence that – in specific contexts and with the aid of concrete materials – even 6-year-old preschoolers already spontaneously discriminate between linear and exponential growth processes when asked to make estimations. Their empirical research suggests the existence of a sensitivity for both linear and non-linear growth in these children, which develops independently from formal mathematics education. However, it remains unclear from which experiential domain young children get their informal knowledge about non-linear growth processes and how this knowledge further develops and interferes with mathematical concepts that are more explicitly taught at school (such as multiplication and proportionality). Evidently, knowledge about the origin and development of the concept of non-linearity could provide important additional information for the design of appropriate teaching methods and could help to improve students' deficient performance on non-linear tasks (see below).

Third, although we have shown that classroom experiences have an indisputable role in the development of students' overuse of linearity, it remains unclear how this development occurs, and to what extent various cases of unwarranted proportional reasoning can be avoided by adaptations in the instructional environment. New design studies are needed to investigate whether long-term interventions including revisions of important parts of the mathematics curriculum on the basis of model-eliciting activities (see next section) can have a substantial impact on students' proportional reasoning. More generally, these design studies could also shed some light on the potential of mathematical modelling as a tool for sense-making and critical analysis of issues important in students' lives, their communities, or society (Mukhopadhyay & Greer, 2001).

Fourth, future research on students' over-reliance on linearity and on their understanding of non-linearity should always take into account the domain-specific mathematical knowledge that students rely on when they approach particular tasks. Although the over-reliance on linearity occurs in various content domains and phenomena may look very similar, they can actually be quite different. Weak or even flawed understandings of certain mathematical concepts – along with the strong relatedness between some mathematical concepts and the idea of linearity – need to be taken into consideration when trying to provide an adequate interpretation of students' reactions, and when developing new intervention studies.

In that respect, it also seems worthwhile to further investigate the characteristics of students' overuse of linearity in other subject-matter domains, especially in science education. History provides many cases of thinkers who inadequately postulated linear relations to grasp situations in physics. At the beginning of chapter 1, we described Aristotle's belief that an object 10 times as heavy as an other object would fall 10 times as fast as that other object, when released from the same height (and it took many centuries before he was proven wrong by Galileo Galilei) (Galilei, 1638). But also, nowadays, students are often reported to have an Aristotelian view on physics (Champagne, Klopfer, & Anderson, 1980; Ebison, 1993; Viennot, 1979; Weller, 1995). Anderson (1983, p. 242) has reported that "subjects have a tendency to assume and impose linear relations". A more systematic inventory and analysis of the various manifestations of students' overuse of linearity in diverse scientific contexts – in a way similar to what we have done in this monograph for some mathematical contexts – could provide a better understanding of the actual impact of the overuse of linearity on students' reasoning and of its educational and psychological roots.

3. RECOMMENDATIONS FOR THE IMPROVEMENT OF EDUCATIONAL PRACTICE

Based on the empirical investigations reported in this monograph, we finally present suggestions for improving the quality of teaching and learning of linear and non-linear modelling at the elementary and secondary school level. First, we suggest a number of concrete modifications of current instructional practice that would constitute an important step forward in counteracting students' tendency towards improper proportional reasoning. Although many of these modifications can be realised by users of most currently used curricula and textbooks for school mathematics, we suggest, in a second section, how they can fit in a more comprehensive reform, which consists in bringing the modelling perspective more to the forefront of mathematical education.

3.1 Concrete suggestions for educational practice

Already at the elementary level, an important goal is to improve in numerous ways the quality of word problems (particularly proportional reasoning tasks) and the way teachers handle this kind of problem. Of course, textbook series worldwide have certainly been improved in this direction, but further progress still seems possible. In this respect, Verschaffel, Greer, and De Corte (2002, pp. 270–271) have made suggestions including the following:

- Break up the expectation that any word problem can be solved by adding, subtracting, multiplying, or dividing, or by a simple combination thereof. An illustrative example in that respect is the 'non-linear' Handshakes problem ("How many handshakes will occur at a party if every one of the n guests shake hands with each of the others?", answer:
  $$\frac{n\,(n-1)}{2};$$
  see, e.g., National Council of Teachers of Mathematics, 1989; Stacey, 1989). Although the Handshakes problem involves a rather complex non-linear mathematical structure, even middle-school students can approach the problem, e.g., by exploring (cooperatively) easier versions of the problem (e.g., for $n = 3, 4, \ldots$), making drawings or other schematizations and looking for patterns, etc.
- Eliminate the flaws in textbooks that allow superficial solution strategies to be undeservedly successful. If a specific context or presentational structure (e.g. the 'missing-value' format) tends to elicit a specific problem-solving routine, this behaviour can be questioned by confronting students with mathematically different problems within the same context and/or presentational structure, and with mathematically identical problems presented in a different context and/or presentational structure. This kind of 'discrimination training using counterexamples' (Greer, 2006) is especially recommendable for the context or presentational structure in which students encounter a certain mathematical model for the first time.
- Vary problems so that it cannot be assumed that all data included in the problem, and only those data, are required for solution. As in real life, students must learn to select the data that are needed to solve a problem,

and eventually look for 'missing' data. In this respect, one could think about project-like tasks embedded in a rich context in which lots of data are given or can be calculated, measured or estimated for different problem-solving purposes (such as the integrative project of our teaching experiment reported in chapter 5).

- Weed out word problems in which the situation, the numbers, and/or the question do not correspond to real life or for which the mathematical model that students are expected to find and apply does not fit (well) with the situation evoked by the problem statement. For instance, we once encountered a textbook problem wherein a proportional relation between the weight of laundry and the washing time was assumed… As shown by Verschaffel et al. (2000), students implicitly learn from such tasks to put reality between brackets in the mathematics classroom!
- Legitimize forms of answer other than exact numerical answers, e.g. estimations, commentaries, drawings, graphs, etc. This recommendation fits with the plea of many math educators to emphasise the process rather than the result of a problem-solving activity. It can also stimulate students to increase their repertoire of problem-solving strategies and approaches.
- Include alternative forms of tasks such as classification tasks and problem-posing tasks. In a classification task, students are asked to group problems in different categories and to explain the motivation for their grouping (e.g., De Bock, Van Dooren, & Verschaffel, 2005). Problem posing means that teachers create opportunities for children to generate problems themselves, complementary to giving them experience in problem solving (e.g., Cai & Hwang, 2002; Ellerton, 1986; Silver, 1994; Silver & Cai, 1996; Silver, Shapiro, & Deutsch, 1993; Verschaffel, De Corte, Lowyck, Dhert, & Vandeput, 2000). Such alternative tasks can move the attention from individually calculating numerical answers towards classroom discussions on the link between problem situations and arithmetical operations.

When focusing especially on the teaching and learning of proportionality, elementary school textbooks should contain more variation in proportional reasoning tasks and take care that these tasks are not always presented in a missing-value format. Moreover, varied tasks explicitly focussing on distinguishing proportional and non-proportional models and reasoning should be included in textbooks. As reflected in the *Principles and standards for school mathematics* (National Council of Teachers of Mathematics, 2000, p. 217),

> facility with proportionality involves much more than setting two ratios equal and solving for the missing term. It involves recognising quantities

that are related proportionally and using numbers, tables, graphs, and equations to think about the quantities and their relationship.

Bringing more variation in problem situations and formulations should avoid the proportional scheme being automatically triggered by a stereotypical linguistic format. The way students inductively learn from examples – and hence the importance of carefully choosing the examples to which students are exposed – was already excellently analysed and discussed by behavioural psychologists in the twenties of the previous century (see, e.g., Thorndike et al., 1924) and 'exemplification' (i.e. the use of examples in teaching and learning) was and still is a central research focus in the mathematics education community (Bills et al., 2006).

Several examples of proportional reasoning tasks that are stated in alternative ways can be found in the literature on proportional reasoning (see, e.g., Cramer & Post, 1993; National Council of Teachers of Mathematics, 2000; Schwartz & Moore, 1998) and on task design (Ainley & Pratt, 2005). Some examples are:

- You are the ace detective for a local law enforcement agency. You have been called in to help to solve a case. Suspects have been narrowed to 3 people. One is 5 feet tall. The second is 6 feet tall. The third is 7 feet tall. The only clue left from the scene of the crime is a handprint. Use the handprint to help narrow the list of suspects. [A handprint that had a span approximately one-third greater than that of a typical adult who is 6 feet tall is shown with the problem.] (Thompson & Bush, 2003).
- Early this morning, the police discovered that, sometime late last night, some nice people rebuilt the old brick drinking fountain in the park. The mayor would like to thank the people who did it. However, nobody saw who it was. All the police could find were lots of footprints. [A box showing one footprint is shown.] The person who made this footprint seems to be very big. Yet, to find this person and his or her friends, it would help if we could figure out how big the person really is. Your job is to make a 'how to' tool kit that the police can use to figure out how big people are – just by looking at their footprints. Your tool kit should work for footprints like the one that is shown. However, it also should work for other footprints. (Lesh & Harel, 2003).
- Begin with a group of three students joining hands in a circle. Record the length of time needed to pass a hand squeeze around the group. A second trial should be conducted in the reverse order and the average of the two trials used. Add an additional group of three students and repeat. Continue to add groups of three until everyone is included. Record the different times and number of students for each time. Allowing for some inaccuracy in measurement, are the ratios of time to number of students

equivalent? Does this case represent a proportional situation? (Nebraska Educational Television Network, 1997).

- In a newspaper interview, prince Filip of Belgium declared himself "optimistic about Chinese economic growth rate of 14% per year". He stated: "It means that people's income will double in the next 7 years…" (Huylebrouck, 2005, p. 30). Comment on this statement.

Figure 7-1. "Last week, this mountain seemed twice as high"… (De Standaard, March 8, 2004; photographed by Christophe Gilbert)

- In 2004, 'De Standaard', a main Flemish newspaper switched over on the so-called tabloid-size. Compared with formerly, the area of the newspaper pages was halved (and the newspaper itself became twice as thick!). To promote its new look, advertisements in the first issue of the newspaper in tabloid-size were all referring to the new size. One of them was showing a picture with two alpinists on a big mountain (see Figure 7-1). The translation of the text on that picture is as follows: "Strange, in the newspaper of last week, this mountain seemed twice as high". Do you think this statement is mathematically correct and, if necessary, can you adapt the text to make it correct?

As evidenced by the results of most studies included in this monograph, students' tendency to over-generalise mathematical models is not limited to the overgeneralisation of the linear model. Once students have discovered the non-linear character of certain situations, many start to over-generalise these newly attained non-linear insights to 'simple' linear situations. This finding is important for the teaching and learning of non-linearity. Teachers, especially at the secondary level, should prevent and/or counteract this overgeneralisation risk by discussing more thoroughly the situations in which these non-linear models are applicable and, already at the beginning of the teaching/learning process, contrasting them with the situations for which the familiar linear models are suitable. As illustrated and recommended by Fischbein (1987) and others, major guiding principles for the development of these instructional practices are: identifying students' intuitive criteria for choosing between linear and non-linear models and discussing their origins, raising awareness of inconsistencies within the students' thinking, and emphasizing the inherent relativity of mathematical models. Historical examples can be quoted to illuminate discussion of the nature of cognitive obstacles and the cognitive changes to overcome them (e.g. Aristotle's and the Méré's linear assumptions in the early development of, respectively, the theories of kinematics and probability and how their linear approaches were rejected by other scholars who looked at the phenomena under investigation from a completely different perspective).

Finally, the research reported in this monograph points to the failure throughout typical mathematics education to adequately address another 'big idea' (Kaput, 1995; Romberg & Kaput, 1999), namely dimensionality, which is crucial to many parts of mathematics and science. The relative size of one-, two-, and three-dimensional elements in similar figures. As stated by Freudenthal (1983), formulas for perimeter and area of the circle and for area and volume of the sphere are didactically and practically overshadowed by the knowledge about their behaviour under enlargement and reduction, which applies in a large field not covered by formulas. More specifically, several studies reported in the central chapters of this monograph revealed students' struggling with the concepts of area and volume and the behaviour of these magnitudes, particularly when dealing with irregular figures. We endorse the frequently-heard warnings against narrowing the teaching and learning of the area and volume to the application of formulas for calculating areas or volumes of regular figures or solids. The main focus should be on conceptual and dimensional thinking. What are different dimensions of a geometrical figure and how do they distinctively behave under operations of enlargement and reduction? This emphasis could be supported didactically by looking at these operations from multiple representational perspectives (graphs, formulas, drawings, and tables), by checking formulas and calculations on their dimensional correctness, and by including activities involving the estimation of areas and volumes of similarly enlarged or reduced shapes, particularly irregular ones. Also, to provide a nuanced and even corrective view, the limitations of straightforwardly applying the mathematical ideas of dimensionality and similarity on physical and biological realities should be discussed (Haldane, 1928; Thompson, 1961; see also chapter 1).

3.2 A more comprehensive instructional approach

Although many of the above-mentioned recommendations can be realised by users of current curricula and textbooks, they actually fit within a more general reform approach, namely to teach mathematics from a genuine modelling perspective (Blum et al., 2002; Lesh & Lehrer, 2003; National Council of Teachers of Mathematics, 1989, 2000; Verschaffel et al., 2000), and to do this already from the very beginning of primary education on (see, e.g., Centre de Recherche sur l' Enseignement des Mathématiques, 2002; Greer, 1987, 2006; Kaput, 1994; Usiskin, 2004; Verschaffel, 2002). Already while teaching basic concepts like addition, substraction, multiplication, division, and direct and inverse proportionality, immediate emphasis should be laid on these concepts' capacity of modelling some situations (at some level of precision) and their inadequacy of modelling others (Usiskin, 2004, 2006). In fact, although mathematical modelling is generally associated with courses at the tertiary or, to an increasing extent, secondary level of instruction, an early exposure to essential modelling ideas, by re-conceptualising the basic arithmetic operations and other primary school content as modelling exercises, can provide a solid base for competently applying mathematics at the primary school level and for further extensions of these mathematical tools found in algebra, geometry, calculus, and statistics at the secondary and tertiary level. Prototypically clean situations can be used to develop students' ability to easily recognize mathematical models and to fluently apply related procedures (or strategies), but at regular times, they should be alternated with exercises in relating more authentic real-world situations to these mathematical models and in reflecting on this relation (Greer, 2006; Verschaffel, 2002; Verschaffel et al., 2000) as a corrective to an oversimplistic view of the world that many *supposed* applications of mathematics tend to establish.

In the last decade, many scholars have argued for the use of so-called model-eliciting activities wherein students are stimulated to invent, extend, revise, and refine many of the important mathematical ideas throughout the mathematics curriculum, such as proportionality (see, e.g., Lesh & Doerr, 2003). Compared to traditional word problems which require students to give short answers to artificially constrained questions in premathematized situations, model-eliciting activities require students to develop the intended conceptual ideas and tools by going through a series of modelling cycles in which the givens, goals, and relevant solutions are continuously reinterpreted, rethought, and renegotiated.

A genuine modelling perspective implies that all phases of the modelling process are considered equally important and receive ample attention. Many authors have proposed descriptions of this process (e.g., Blum & Niss, 1991; Burkhardt, 1994; Mason, 2001; Verschaffel et al., 2000), but, essentially, they all involve the following stages (see Verschaffel et al., 2000) (Figure 7-2):

1. understanding the phenomenon under investigation, leading to a model of the relevant elements, relations and conditions that are embedded in the situation (situation model),
2. constructing a mathematical model of the relevant elements, relations and conditions available in the situation model,
3. working through the mathematical model using disciplinary methods in order to derive some mathematical results,
4. interpreting the outcome of the computational work to arrive at a solution to the real-world problem situation that gave rise to the mathematical model,
5. evaluating the model by checking if the interpreted mathematical outcome is appropriate and reasonable for the original problem situation and
6. communicating the solution of the original real-world problem.

Verschaffel et al. (2000) also provided an elaborated version of this simple six-stage model (Figure 7-3), embodying a more in-depth analysis of the nature of modelling, particularly as a collective activity, wherein

1. knowledge about the real phenomena is not suppressed but considered as a valuable component in the initial stage of the solution process,
2. the nature of the modelling act is influenced by the goals implicit in the situation, imposed by the teacher, or negotiated,
3. the solver can make use of a rich variety of resources (including software modelling tools) in the stages of mathematical modelling and analysis,
4. the interpretation involves comparison of alternative models, and
5. the communication phase may go far beyond a bald reporting of the result of the calculation (for more details, see Verschaffel et al. 2000).

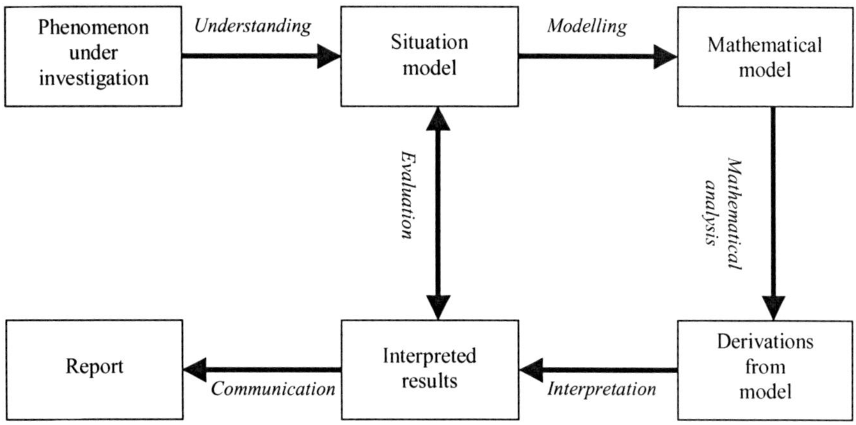

Figure 7-2. Schematic diagram of the process of mathematical modelling (Verschaffel et al., 2000, p. xii)

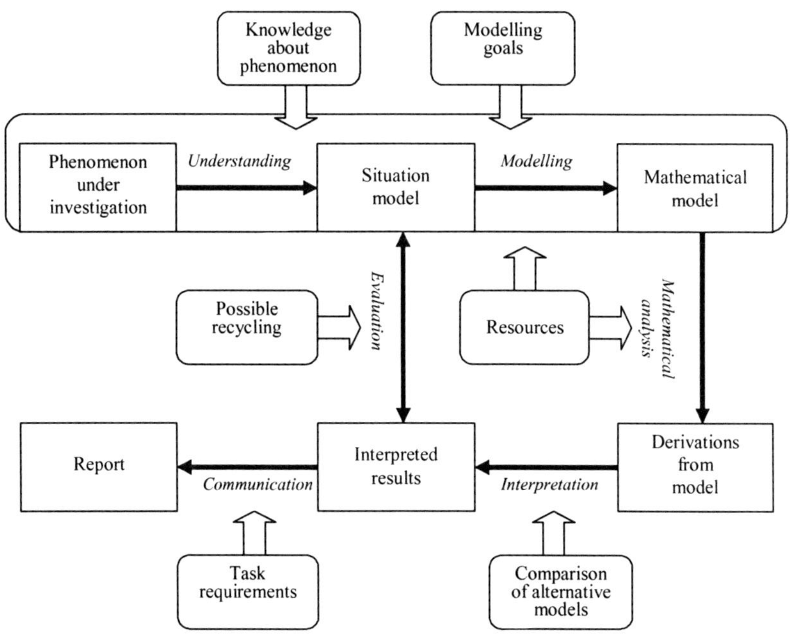

Figure 7-3. Elaborated view of the modelling process (Verschaffel et al., 2000, p. 168)

In recent years, several researchers have set up design studies in which they developed, implemented, and evaluated experimental programs aimed at developing in elementary and middle school students a genuine disposition towards mathematical modelling. Also the teaching experiment reported in chapter 5 was an attempt to strongly integrate the modelling perspective with the domain of geometrical enlargements and reductions, but unfortunately, for many students it did not fulfill all expectations (for reasons that were explained in the discussion session of that chapter). Typical other examples are:

- several developmental research projects from the Freudenthal Institute in The Netherlands (see, e.g., Gravemeijer, 1997),
- the Jasper studies of the Cognition and Technology Group at Vanderbilt (1997) wherein mathematical problem solving is anchored in realistic contexts using new information technologies,
- Lehrer and Schauble's (2000) experimental curriculum for mathematics and science teaching in young children built upon the modelling approach,
- English and Lesh' (2003) recent work around 'ends-in-view' problems wherein neither the givens, nor the goal or the necessary solution steps are specified clearly, and
- the learning environment for mathematical modelling and word problem solving in upper elementary school children developed by Verschaffel, De Corte, Lasure, Van Vaerenbergh, Bogaerts, and Ratinckx (1999).

The following characteristics are common to these experimental programs:

- The introduction of more realistic and challenging tasks, which do involve some, if not most, of the complexities of real modelling tasks (such as the necessity to formulate the problem, to seek and apply aspects of the real context, to select tools to be used, to discuss alternative hypotheses and rival models, to decide upon the level of precision, to interpret and evaluate outcomes).
- Accentuation of multiple representations (including graphs, tables, schemes, drawings, formulas, and words) and their reciprocal relationships to enhance conceptual understanding and deep-level learning.
- A variety of teaching methods and learner activities, including expert modelling of strategic aspects of the modelling process, small-group work and whole-class discussions. Typically the focus is not on presenting and rehearsing established mathematical models, but rather on demonstrating, experiencing, articulating, and discussing what modelling is all about.
- Creating a classroom climate that is conducive to the development of genuine mathematical modelling and of the accompanying beliefs.

Generally, these studies have produced moderately positive outcomes in terms of performance, underlying processes, and motivational and affective aspects of learning. After reviewing the available research evidence, Niss (2001, p. 8) concluded that

> application and modelling capability can be learnt – and according to the above-mentioned findings has to be learnt – but at a cost, in terms of effort, complexity of task, time consumption, and reduction of syllabus in the traditional sense.

To some extent, the above-mentioned characteristics of the modelling approach are beginning to be implemented in mathematical frameworks and tests in many countries. However, according to Niss (2001) and Burkhardt (2004), in general international terms, genuine and extensive applications and modelling perspectives and activities continue to be scarce in everyday practices of mathematics education. Niss points to two important barriers, namely the difficulty of getting the modelling perspective into tests (in which it is easier to include tasks with a more closed character), and the extremely high demands that such a modelling approach puts on teachers (mathematically, pedagogically, and personally).

A final important issue to be addressed is whether the teaching and learning of mathematical applications according to the modelling perspective is important, and feasible for, all students. Over the past few years several authors, such as Blum and Niss (1991), Keitel (1989), Lesh and Doerr (2003) and Mukhopadhyay and Greer (2001), have made strong pleas for engaging *all* students in the modelling perspective, both for the empowerment of the individual and for the betterment of society. In relation to this 'political aspect', the most important reason to for introducing the modelling perspective to all students is to help as many people as possible

> to become critical thinkers who can use mathematics as a tool for analyzing social and political issues, and can reflect on that tool use, including its limitations (Mukhopadhyay & Greer, 2001, p. 310).

Proportionality is a key concept for this kind of analysis because it is often taken as a baseline model in using data to test for fairness in many contexts (e.g. how does the percentage of people in a certain group selected for employment, university, punishment, etc. relate to the percentage of that group in the population as a whole?). In short, proportionality is an essential concept in the mathematical analysis of equity (see also Gutstein, 2003). Quantitative analyses of social and political realities are authentic modelling activities (in the above-mentioned sense), also because there usually are extenuating and complicating circumstances which need to be taken into account in evaluating the reasonableness of the applied model or deviations therefrom.

Evidently, once mathematics educators start applying this modelling perspective on a larger scale and allow students to bring in their personal experience when trying to make sense of all kinds of technical, social, political, and cultural issues and phenomena, they will be confronted quickly and inevitably with the diversity of these experiences in terms of gender, social class, and ethnic diversity (see also, Boaler, 1994; Cooper & Dunne, 1998; Tate, 1994). We endorse Mukhopadhyay and Greer's (2001) claim that engaging students in such modelling activities, with careful attention to the relevance of the problem contexts and all the diversity in views and approaches that they elicit, may be the best way to prevent students from becoming alienated by mathematics and its authority, and to help them using mathematics as a powerful personal tool for the analysis of issues important in their personal lives and in society. An illustrative example in this respect, only using 'simple' proportional reasoning, is the analysis of the phenomenon of 'Barbie dolls' described in Mukhopadhyay and Greer (2001). Mukhopadhyay has developed several activities in this Barbie context (e.g. with respect to body image and body type in relation to self-esteem and cultural identity, labour conditions of overseas factory workers for American companies, etc.) in order to enhance students' disposition towards critical thinking. According to Mukhopadhyay and Greer (2001; see also Greer, Verschaffel, & Mukhopadhyay, 2006), it is important and also feasible to start applying the modelling perspective seriously in the mathematics education of all students from a (very) young age on and with a diversity of learners.

References

Ainley, J., & Pratt, D. (2005). The significance of task design in mathematics education: Examples from proportional reasoning. In H. L. Chick & J. L. Vincent (Eds.), *Proceedings of the 29th Conference of the International Group for the Psychology of Mathematics Education* (Vol. 1, pp. 93–122). Melbourne, Australia: University of Melbourne.

Ainsworth, S., Bibby, P., & Wood, D. (2002). Examining the effects of different multiple representational systems in learning primary mathematics. *The Journal of the Learning Sciences, 11*(1), 25–61.

Akatufba, A. H., & Wallace, J. (1999). Mathematical dimensions of students' use of proportional reasoning in high school physics. *School Science & Mathematics, 99*(1), 31–41.

Anderson, N. H. (1983). Intuitive physics: Understanding and learning of physical relations. In T. J. Tighe & B. E. Shepp (Eds.), *Perception, cognition and development: Interactional analyses* (pp. 231–265). Mahwah NJ: Lawrence Erlbaum Associates.

Anderson, N. H., & Cuneo, D. (1978). The height and width rule in children's judgements of quantity. *Journal of Experimental Psychology: General, 107*, 335–378.

Aprea, C., & Ebner, H. J. G. (1999, August). *The impact of active graphical representation on the acquisition and application of knowledge in the context of business education.* Paper presented at the 8th European Conference for Research on Learning and Instruction, Göteborg, Sweden.

Behr, M. J., Harel, G., Post, T., & Lesh, R. (1992). Rational number, ratio, and proportion. In D. A. Grouws (Ed.), *Handbook of research on mathematics thinking and learning* (pp. 296–333). New York: Macmillan.

Ben-Zeev, T., & Star, J. R. (2001). Spurious correlations in mathematical thinking. *Cognition and Instruction, 19*, 253–275.

Berté, A. (Réd.). (1987; 2nde édition, 1992). *Enseignement des mathématiques utilisant la 'réalité' (Tome 1)* [Teaching mathematics using 'reality' (Part 1)]. Bordeaux: IREM.

Berté, A. (1993). *Mathématique dynamique* [Dynamical mathematics]. Paris: Nathan.

Bills, L., Mason, J., Watson, A., Zaslavsky, O., Goldenberg, P., Rowland, T., & Zazkis, R. (2006). Exemplification: The use of examples in teaching and learning mathematics. In J. Novotná, H. Moraová, M. Krátká, & N. Stehlíková (Eds.), *Proceedings of the 30th Conference of the International Group for the Psychology of Mathematics Education* (Vol. 1, pp. 125–154). Czech Republic, Prague.

Blum, W., & Niss, M. (1991). Applied mathematical problem solving, modelling, applications, and links to other subjects – State, trends, and issues in mathematics education. *Educational Studies in Mathematics, 22*, 37–68.

Blum, W., et al. (2002). ICMI Study 14: Applications and modelling in mathematics education – Discussion document. *Educational Studies in Mathematics, 51*, 149–171.

Boaler, J. (1993). The role of contexts in the mathematics classroom: Do they make mathematics more real? *For the Learning of Mathematics, 13*(2), 12–17.

Boaler, J. (1994). When do girls prefer football to fashion? An analysis of female underachievement in relation to 'realistic' mathematic contexts. *British Educational Research Journal*, *20*(5), 551–564.

Boaler, J. (2000). Exploring situated insights into research and learning. *Journal for Research in Mathematics Education, 39*(1), 113–119.

Bold, B. (1969). *Famous problems of geometry and how to solve them*. New York: Dover Publications.

Brousseau, G. (1997). *Theory of didactical situations in mathematics* (Edited and translated by N. Balacheff, M. Cooper, R. Sutherland, & V. Warfield). Dordrecht, The Netherlands: Kluwer.

Burkhardt, H. (1994). Mathematical applications in school curriculum. In T. Husen & T. N. Postlethwaite (Eds.), *The international encyclopedia of education* (2nd ed., pp. 3 631–3 634). Oxford, England: Pergamon.

Burkhardt, H. (2004). Establishing modelling in the curriculum: Barriers and levers. In H.-W. Henn & W. Blum (Eds.), *ICMI Study 14: Applications and modelling in mathematics education – Pre-conference volume* (pp. 53–58). Dortmund, Germany.

Cai, J., & Hwang, S. (2002). Generalized and generative thinking in U.S. and Chinese students' mathematical problem solving and problem posing. *Journal of Mathematical Behavior, 21,* 401–421.

Centre de Recherche sur l'Enseignement des Mathématiques. (2002). *Des grandeurs aux espaces vectoriels: La linéarité comme fil conducteur* [From quantities to vector spaces: Linearity as a general guiding principle]. Nivelles: Author.

Champagne, A. B., Klopfer, L. E., & Anderson, J. H. (1980). Factors influencing the learning of classical mechanics. *American Journal of Physics, 48*, 1074–1079.

Claes, K. (2000). *De lineariteitsillusie doorbroken? Een empirisch onderzoek naar het effect van de authenticiteit van de context en van tekeningen op het oplossen van niet-lineaire toepassingsproblemen bij leerlingen van het secundair onderwijs* [The illusion of linearity defeated? An empirical research on the effect of context authenticity and of

drawings on secondary school students' solution of non-linear application problems]. Unpublished master's thesis, Center for Instructional Psychology and Technology, University of Leuven, Belgium.
Clarkson, R. (1989). Ratio: Enlargement. In D. C. Johnson (Ed.). *Children's mathematical frameworks 8–13: A study of classroom teaching* (pp. 191–217). Windsor, U.K.: Nelson.
Cobb, P. (1994). Where is the mind? Constructivist and sociocultural perspectives on mathematical development. *Educational Researcher, 23*(7) 13–20.
Cobb, P., Yackel, E., & McClain, K. (2000). *Symbolizing and communicating in mathematics classrooms*. Mahwah, NJ: Lawrence Erlbaum Associates.
Cobb, P., Yackel, E., & Wood, T. (1992). A constructivist alternative to the representational view of mind in mathematics education. *Journal for Research in Mathematics Education, 23*, 2–33.
Cognition and Technology Group at Vanderbilt. (1997). *The Jasper project: Lessons in curriculum, instruction, assessment, and professional development.* Mahwah, NJ: Lawrence Erlbaum Associates.
Collins, A., Brown, J. S., & Newman, S. E. (1989). Cognitive apprenticeship: Teaching the crafts of reading, writing and mathematics. In L. B. Resnick (Ed.), *Knowing, learning and instruction. Essays in honor of Robert Glaser* (pp. 453–494). Mahwah, NJ: Lawrence Erlbaum Associates.
Cooper, B. (1994). Authentic testing in mathematics? The boundary between everyday and mathematical knowledge in National Curriculum testing in English schools. *Assessment in Education, 1*, 143–166.
Cooper, B., & Dunne, M. (1998). Anyone for tennis? Social class differences in children's responses to National Curriculum mathematics testing. *Sociological Review, 46*, 115–146.
Cooper, B., & Harries, T. (2002). Children's responses to contrasting 'realistic' mathematics problems: Just how realistic are children ready to be? *Educational Studies in Mathematics, 49*, 1–23.
Cooper, B., & Harries, T. (2003). Children's use of realistic considerations in problem solving: some English evidence. *Journal of Mathematical Behaviour, 22*, 451–465.
Cramer, K., & Post, T. (1993). Connecting research to teaching: Proportional reasoning. *Mathematics Teacher*, *86*, 404–407.
Cramer, K., Post, T., & Currier, S. (1993). Learning and teaching ratio and proportion: Research implications. In D. T. Owens (Ed.), *Research ideas for the classroom: Middle grades mathematics* (pp. 159–178). New York: Macmillan.
Daumas, D. (1989). La démonstration de l'irrationalité chez les grecs [Demonstration of irrationality by the Greek]. In Commission inter-IREM Histoire et Epistémologie des Mathématiques (Ed.), *La démonstration dans l'histoire* [Demonstration in history] (pp. 389–423). Lyon: IREM.
Davis, R. B. (1989). The culture of mathematics and the culture of schools. *Journal of Mathematical Behavior*, *8*, 143–160.
De Block-Docq, C. (1992). *Analyse épistémologique comparative de deux enseignements de la géométrie plane vers l'age de douze ans* [Epistemological comparative analysis of two approaches to plane

geometry on the age of twelve]. Unpublished doctoral dissertation, Faculté des Sciences, Université Catholique de Louvain, Louvain-la-Neuve, Belgium.

De Bock, D., Van Dooren, W., Janssens, D., & Verschaffel, L. (2002a). Improper use of linear reasoning: An in-depth study of the nature and the irresistibility of secondary school students' errors. *Educational Studies in Mathematics, 50*(3), 311–334.

De Bock, D., Van Dooren, W., & Verschaffel, L. (2005). Not everything is proportional: Task design and small-scale experiment. In H. L. Chick & J. L. Vincent (Eds.), *Proceedings of the 29th Conference of the International Group for the Psychology of Mathematics Education* (Vol. 1, pp. 97–102). Melbourne, Australia: University of Melbourne.

De Bock, D., Van Dooren, W., Verschaffel, L., & Janssens, D. (2001). Secondary school pupils' improper proportional reasoning: An in-depth study on the nature and persistence of pupils' errors. In M. van den Heuvel-Panhuizen (Ed.), *Proceedings of the 25th Conference of the International Group for the Psychology of Mathematics Education* (Vol. 2, pp. 321–328). Utrecht, The Netherlands: Utrecht University.

De Bock, D., Verschaffel, L., & Janssens, D. (1996). *De lineariteitsillusie: Een exploratief onderzoek* [The illusion of linearity: An explorative study]. Internal report, Center for Instructional Psychology and Technology, University of Leuven, Belgium.

De Bock, D., Verschaffel, L., & Janssens, D. (1998). The predominance of the linear model in secondary school students' solutions of word problems involving length and area of similar plane figures. *Educational Studies in Mathematics*, *35*, 65–83.

De Bock, D., Verschaffel, L., & Janssens, D. (2002b). The effects of different problem presentations and formulations on the illusion of linearity in secondary school students. *Mathematical Thinking and Learning, 4*, 65–89.

De Bock, D., Verschaffel, L., Janssens, D., Van Dooren, W., & Claes, K. (2003). Do realistic contexts and graphical representations always have a beneficial impact on students' performance? Negative evidence from a study on modelling non-linear geometry problems. *Learning and Instruction, 13*(4), 441–463.

De Corte, E., Greer, B., & Verschaffel, L. (1996). Psychology of mathematics teaching and learning. In D. C. Berliner & R. C. Calfee (Eds.), *Handbook of educational psychology* (pp. 491–549). New York: Macmillan.

De Corte, E., Verschaffel, L., & Masui, C. (2004). The CLIA-model: A framework for designing powerful learning environments for thinking and problem solving. *European Journal for Psychology of Education, 19*, 365–384.

DeFranco, T. C., & Curcio, F. R. (1997). A division problem with a remainder embedded across two contexts : Children's solutions in restrictive versus real-world settings. *Focus on Learning Problems in Mathematics, 19*(2), 58–72.

Dehaene, S. (1997). *The number sense: How the mind creates mathematics.* New York: Oxford University Press.

de Lange, J. (1987). *Mathematics, insight and meaning. Teaching, learning and testing of mathematics for the life and social sciences.* Utrecht: Vakgroep Onderzoek Wiskundeonderwijs en Onderwijscomputercentrum.
Dirkes, M. A. (1991). Draw to understand. *Arithmetic Teacher, 39*(4), 26–29.
Ebersbach, M., & Resing, W. C. M. (2006). Reasoning about non-linearity in 6- to 9-year-olds: The role of task presentation. In J. Novotná, H. Moraová, M. Krátká, & N. Stehlíková (Eds.), *Proceedings of the 30th Conference of the International Group for the Psychology of Mathematics Education* (Vol. 3, pp. 9–16). Czech Republic, Prague.
Ebison, M. G. (1993). Newtonian in mind, but Aristotelian at heart. *Science and Education, 2*, 345–362.
Ellerton, N. F. (1986). Children's made-up mathematical problems: a new perspective on talented mathematicians. *Educational Studies in Mathematics, 17,* 261–271.
English, L., & Lesh, R. (2003). Ends-in-view problems. In R. Lesh & H. Doerr (Eds.), *Beyond constructivism. Models and modeling perspectives on mathematical problem solving, learning and teaching* (pp. 297–316). Mahwah, NJ: Lawrence Erlbaum Associates.
Esteley, C., Villarreal, M., & Alagia, H. (2004). Extending linear models to non-linear contexts: An in-depth study about two university students' mathematical productions. In M. J. Høines & A. B. Fuglestad (Eds.), *Proceedings of the 28th Conference of the International Group for the Psychology of Mathematics Education* (Vol. 2, pp. 343–350). Bergen, Norway.
Evangelidou, A., Spyrou, P., Elia, I., & Gagatsis, A. (2004). University students' conceptions of functions. In M. J. Høines & A. B. Fuglestad (Eds.), *Proceedings of the 28th Conference of the International Group for the Psychology of Mathematics Education* (Vol. 2, pp. 351–358). Bergen, Norway.
Evans, J. S. B. T. (2003). In two minds: Dual process accounts of reasoning. *Trends in Cognitive Sciences, 7*, 454–459.
Feys, R. (1995). Meten en metend rekenen [Measurement and metrication]. In L. Verschaffel & E. De Corte (Eds.), *Naar een nieuwe reken/wiskundedidactiek voor de basisschool en de basiseducatie. Deel 3: Verder bouwen aan gecijferdheid* [Towards a new didactic of arithmetic and mathematics for the elementary school and the basic education. Part 3: Building further on numeracy] (pp. 99–135). Brussel/Leuven: Studiecentrum Open Hoger Onderwijs (StOHO)/Acco.
Fischbein, E. (1987). *Intuition in science and mathematics.* Dordrecht: Reidel.
Fischbein, E. (1999). Intuitions and schemata in mathematical reasoning. *Educational Studies in Mathematics, 38*, 11–50.
Fischbein, E., & Gazit, A. (1984). Does the teaching of probability improve probabilistic intuitions? *Educational Studies in Mathematics, 15*, 1–24.
Fischbein, E., & Schnarch, D. (1997). The evolution with age of probabilistic, intuitively based misconceptions. *Journal for Research in Mathematics Education, 28*, 96–105.
Forman, E. A., & Cazden, C. B. (1985). Exploring Vygotskian perspectives in education. The cognitive value of peer interaction. In J. V. Wertsch

(Ed.), *Culture, communication, and cognition: Vygotskian perspectives* (pp. 323–347). New York: Cambridge University Press.

Freudenthal, H. (1973). *Mathematics as an educational task*. Dordrecht: Reidel.

Freudenthal, H. (1978). *Weeding and sowing*. Dordrecht: Reidel.

Freudenthal, H. (1983). *Didactical phenomenology of mathematical structures*. Dordrecht: Reidel.

Freudenthal, H. (1991). *Revisiting mathematics education*. Dordrecht: Kluwer.

Gagatsis, A., & Kyriakides, L. (2000). Teachers' attitudes towards their pupils' mathematical errors. *Educational Research and Evaluation, 6*(1), 24–58.

Galilei, G. (1638). *Discorsi e dimostrazioni matematiche intorna a due nuove scienze: 1954, Dialogues concerning two new sciences*. New York: Dover.

Ginsburg, H. P., Kossan, N. E., Schwartz, R., & Swanson, D. (1982). Protocol methods in research on mathematical thinking. In H. P. Ginsburg (Ed.), *The development of mathematical thinking* (pp. 7–47). New York: Academic Press.

Gravemeijer, K. (1994). *Developing realistic mathematics education*. Utrecht, The Netherlands: Freudenthal Institute.

Gravemeijer, K. (1997). Solving word problems: a case of modelling? *Learning and Instruction, 7,* 389–397.

Greer, B. (1987). Understanding of arithmetical operations as models of situations. In J. A. Sloboda & D. Rogers (Eds.), *Cognitive Processes in Mathematics* (pp. 60–80). Oxford: Clarendon.

Greer, B. (1993). The mathematical modelling perspective on wor(l)d problems. *Journal of Mathematical Behavior, 12*, 239–250.

Greer, B. (1997). Modelling reality in the classrooms: The case of word problems. *Learning and Instruction*, *7*(4), 293–307.

Greer, B. (2006). Designing for conceptual change. In J. Novotná, H. Moraová, M. Krátká, & N. Stehlíková (Eds.), *Proceedings of the 30th Conference of the International Group for the Psychology of Mathematics Education* (Vol. 1, pp. 175–178). Czech Republic, Prague.

Greer, B., Verschaffel, L., & Mukhopadhyay, S. (2006). Modelling for life: Mathematics and children's experience. In W. Blum, P. Galbraith, H.-W. Henn, & M. Niss (Eds.). *Applications and Modelling in Mathematics Education. New ICMI Studies Series no. 10*. New York: Springer.

Gutstein, E. (2003). Teaching and learning mathematics for social justice in an urban, Latino school. *Journal for Research in Mathematics Education, 34,* 37–73.

Haldane, J. B. S. (1928). *On being the right size in possible worlds*. New York: Books for Libraries Press.

Hall, V. C., Bailey, J., & Tillman, C. (1997). Can student-generated illustrations be worth ten thousand words. *Journal of Educational Psychology*, *89*(4), 677–681.

Hart, K. M. (1981). *Children's understanding of mathematics: 11–16*. London: Murray.

Hatano, G. (1988). Social and motivational bases for mathematical understanding. *New Directions for Child Development, 41*, 55–70.

Hawkins, A. S., & Kapadia, R. (1984). Children's conceptions of probability – A psychological and pedagogical review. *Educational Studies in Mathematics, 15*, 349–377.

Hinsley, D., Hayes, J., & Simon, H. (1977). From words to equations, meaning and representation in algebra word problems. In M. A. Just & P. A. Carpenter (Eds.), Cognitive processes in comprehension (pp. 89–108). Mahwah, NJ: Lawrence Erlbaum Associates.

Huylebrouck, D. (2005). Bang van Prins Filip [Afraid of prince Filip]. *De Standaard*, June 2.

Kahneman, D. (2002). *Maps of bounded rationality: A perspective on intuitive judgement and choice (Nobel Prize Lecture, December 8)* [On-line]. Available: http://nobelprize.org/economics/laureates/2002/kahnemann-lecture.pdf.

Kahneman, D., & Tversky, A. (1972). Subjective probability: A judgement of representativeness. *Cognitive Psychology, 3*, 430–454.

Kaput, J. J. (1994). Democratizing access to calculus: New routes to old roots. In A. H. Schoenfeld (Ed.), *Mathematical thinking and problem solving* (pp. 77–156). Mahwah, NJ: Lawrence Erlbaum Associates.

Kaput, J. J. (1995). Long-term algebra reform: Democratizing access to big ideas. In C. Lacampagne, W. Blair, & J. J. Kaput (Eds.), The Algebra Colloquium (Vol. 1, pp. 33–49). Washington, DC: US Department of Education.

Kaput, J. J., & West, M. M. (1994). Missing-value proportional reasoning problems: Factors affecting informal reasoning patterns. In G. Harel & J. Confrey (Eds.), *The development of multiplicative reasoning in the learning of mathematics* (pp. 235–287). New York: State University of New York Press.

Karplus, R., Pulos, S., & Stage, E. (1983a). Proportional reasoning of early adolescents. In R. Lesh & M. Landau (Eds.), *Acquisition of mathematical concepts and processes* (pp. 45–89). New York: Academic Press.

Karplus, R., Pulos, S., & Stage, E. K. (1983b). Early adolescents proportional reasoning on 'rate' problems. *Educational Studies in Mathematics, 14*, 219–233.

Kenworthy, D. (Producer), & Sturridge, C. (Director). (1996). *Gulliver's travels* [Film]. (Available from Twentieth Century Fox Home Entertainment, Inc.)

Kindt, M., & de Lange, J. (1986). *Hewet wiskunde. Differentiëren 3* [Hewet mathematics. Differentiating 3]. Culemborg: Educaboek.

Klein, F. (1939). *Elementary mathematics from an advanced standpoint.* New York: Dover.

Konold, C. (1989). Informal conceptions of probability. *Cognition and Instruction, 6*(1), 59–98.

Kuchemann, D., & Hoyles, C. (2001). Investigating factors that influence students' mathematical reasoning. In M. van den Heuvel-Panhuizen (Ed.), *Proceedings of the 25th Conference of the International Group for the Psychology of Mathematics Education* (Vol. 3, pp. 85–92). Utrecht, The Netherlands: Utrecht University.

Kuhn, T. (1962). *The structure of scientific revolutions.* Chicago: University Press.

Lamprianou, I., & Lamprianou, T. A. (2002). The nature of pupils' probabilistic thinking in primary school pupils in Cyprus. In A. D. Cockburn & E. Nardi (Eds.), *Proceedings of the 26th Conference of the International Group for the Psychology of Mathematics Education* (Vol. 3, pp. 273–280). Norwich, U.K.

Landis, J. R., & Koch, G. G. (1977). The measurement of observer agreement for categorical data. *Biometrics, 33*, 159–174.

Larkin, J. H., & Simon, H. A. (1987). Why a diagram is (sometimes) worth ten thousand words. *Cognitive Science*, *12*, 101–138.

Lave, J. (1988). *Cognition in practice*. Cambridge: University Press.

Lave, J. (1992). Word problems: A microcosm of theories of learning. In P. Light & G. Butterworth (Eds.), *Context and cognition. Ways of learning and knowing* (pp. 74–92). New York: Harvester Wheatsheaf.

Lehrer, R., & Schauble, L. (2000). Modeling in mathematics and science. In R. Glaser (Ed.), *Advances in instructional psychology, Volume 5* (pp. 101–159). Mahwah, NJ: Lawrence Erlbaum Associates.

Leinhardt, G., Zaslavsky, O., & Stein, M. K. (1990). Functions, graphs, and graphing: Tasks, learning, and teaching. *Review of Educational Research, 60*, 1–64.

Lelouard, M., Mira, C., & Nicolle, J. M. (1989). Différentes formes de démonstrations dans les mathématiques grecques [Different forms of demonstrations in Greek mathematics]. In Commission inter-IREM Histoire et Epistémologie des Mathématiques (Ed.), *La démonstration dans l'histoire* [Demonstration in history] (pp. 155–180). Lyon: IREM.

Lesh, R. (1992). Assessing authentic mathematical performance. In R. Lesh & S. Lamon (Eds.), *Assesment of authentic performance in school mathematics* (pp. 17–62). Washington, DC: American Associoation for the Advancement of Science.

Lesh, R., & Doerr, H. M. (Eds.) (2003). *Beyond constructivism. Models and modeling perspectives on mathematical problem solving, learning and teaching*. Mahwah, NJ: Lawrence Erlbaum Associates.

Lesh, R., & Harel, G. (2003). Problem solving, modeling, and local conceptual development. *Mathematical Thinking and Learning, 5*(2–3), 157–189.

Lesh, R., & Lehrer, R. (2003). Models and modeling perspectives on the development of students and teachers. *Mathematical Thinking and Learning, 5*, 109–129.

Lesh, R., Post, T., & Behr, M. (1988). Proportional reasoning. In J. Hiebert & M. Behr (Eds.), *Number concepts and operations in the middle grades* (pp. 93–118). Reston, VA: National Council of Teachers of Mathematics.

Limón, M. (2001). On the cognitive conflict as an instructional strategy for conceptual change: A critical appraisal. *Learning and Instruction*, *11*(4–5), 357–380.

Lin, F.-L. (1991). Understanding in multiple ratio and non-linear ratio. *Proceedings of the National Science Council ROC(D), 1*(2), 14–30.

Lin, F.-L., & Yang, K.-L. (2004). Differentiation of students' reasoning on linear and quadratic geometric number patterns. In M. J. Høines & A. B. Fuglestad (Eds.), *Proceedings of the 28th Conference of the International Group for the Psychology of Mathematics Education* (Vol. 4, pp. 457–464). Bergen, Norway.

Linchevski, L., Olivier, A., Sasman, M. C., & Liebenberg, R. (1998). Moments of conflict and moments of conviction in generalising. In A. Olivier & K. Newstead (Eds.), *Proceedings of the 22nd Conference of the International Group for the Psychology of Mathematics Education* (Vol. 3, pp. 215–222). Stellenbosch, South Africa.

Litwiller, B., & Bright, G. (Eds.) (2002). *Making sense of fractions, ratios and proportions. 2002 yearbook.* Reston, VA: National Council of Teachers of Mathematics.

Markovits, Z., Eylon, B.-S., & Bruckheimer, M. (1986). Functions today and yesterday. *For the Learning of Mathematics, 6*(2), 18–24, 28.

Mason, L. (2001). Introduction to special issue "Instructional practices for conceptual change in science domains". *Learning and Instruction, 11*, 259–263.

Matz, M. (1982). Towards a process model for high school algebra errors. In D. Sleeman & J. S. Brown (Eds.), *Intelligent tutoring systems* (pp. 25–50). London: Academic Press.

Ministerie van de Vlaamse Gemeenschap. (1997). *Decreet van juli 1997 tot bekrachtiging van de ontwikkelingsdoelen en eindtermen van het gewoon basisonderwijs* [Educational standards for the regular elementary school]. Brussels: Department for Educational Development, Flemish Community of Belgium.

Ministerie van de Vlaamse Gemeenschap. (2002). *Decreet van januari 2002 tot bekrachtiging van de eindtermen van de tweede en derde graad van het gewoon secundair onderwijs* [Educational standards for the second and third grade of regular secondary school]. Brussels: Department for Educational Development, Flemish Community of Belgium.

Modestou, M., Gagatsis, A., & Pitta-Pantazi, D. (2004). Students' improper proportional reasoning: The case of area and volume of rectangular figures. In M. J. Høines & A. B. Fuglestad (Eds.), *Proceedings of the 28th Conference of the International Group for the Psychology of Mathematics Education* (Vol. 3, pp. 345–352). Bergen, Norway.

Mogensen, P. (2004). A sense of proportions in engineering mathematics. In *Proceedings of the 12th SEFI Maths Working Group Seminar*. Vienna, University of Technology.

Mukhopadhyay, S., & Greer, B. (2001). Modelling with purpose: Mathematics as a critical tool. In B. Atweh, H. Forgasz & B. Nebres (Eds.), *Socio-cultural aspects in mathematics education* (pp. 295–311). Mahwah, NJ: Lawrence Erlbaum Associates.

Mullet, E., Lautrey, J., & Glaser, P. L. (1989). Information integration in an area judgement task: Effect of aids on children's perceptual judgement. *Journal of Genetic Psychology*, *150*, 375–387.

National Council of Teachers of Mathematics. (1989) *Curriculum and evaluation standards for school mathematics*. Reston, VA: Author.

National Council of Teachers of Mathematics. (2000). *Principles and standards for school mathematics*. Reston, VA: Author.

Nebraska Educational Television Network. (1997). *Math vantage unit III*: *Proportional reasoning*. Lincoln: University of Nebraska, Center for Science, Mathematics and Computer Education.

Nesher, P. (1996). School stereotype word problems and the open nature of applications. In C. Alsina, J. M. Alvarez, B. Hodgson, C. Laborde, &

A. Pérez (Eds.), *Selected lectures from the 8th International Congress on Mathematical Education* (pp. 335–343). Sevilla, Spain.

Niss, M. (2001). Issues and problems of research on the teaching and learning of applications and modelling. In J. F. Matos, W. Blum, S. K. Houston, & S. P. Carreira (Eds), *Modelling and mathematics education. ICTMA9: Applications in science and technology* (pp. 72–89). Chichester, U.K.: Horwood.

Noelting, G. (1980a). The development of proportional reasoning and the ratio concept. Part I: The differentiation of stages. *Educational Studies in Mathematics, 11*(3), 217–253.

Noelting, G. (1980b). The development of proportional reasoning and the ratio concept. Part II: Problem structure at successive stages: Problem-solving strategies and the mechanism of adaptive restructuring. *Educational Studies in Mathematics, 11*(4), 331–363.

Nunes, T., Schliemann, A., & Carraher, D. (1993). *Street mathematics and school mathematics*. Cambridge, U.K.: Cambridge University Press.

Orton, J., & Orton, A. (1994). Students' perception and use of pattern and generalization. In J. P. da Ponte & J. F. Matos (Eds.), *Proceedings of the 18th Conference of the International Group for the Psychology of Mathematics Education* (Vol. 3, pp. 407–414). Lisbon, Portugal.

Outhred, L. N., & Mitchelmore, M. C. (2000). Young children's intuitive understanding of rectangular area measurement. *Journal for Research in Mathematics Education, 31*(2), 144–167.

Palm, T. (2002). *The realism of mathematical school tasks. Features and consequences.* Doctoral dissertation, Department of Mathematics, Umeå University, Umeå, Sweden.

Perret-Clermont, A.-N. (1980). *Social interaction and cognitive development in children* (European Monographs in Social Psychology No. 19). New York: Academic Press.

Piaget, J. (1968). *Epistemology and psychology of functions*. Dordrecht: Reidel.

Pólya, G. (1945; 2nd edition, 1957). *How to solve it*. Princeton: Princeton University Press.

Poortvliet, R., & Huygen, W. (1977). *Gnomes*. New York: Harry N. Abrams Inc.

Reed, S. K. (1999). *Word problems: Research and curriculum reform*. Mahwah, NJ: Lawrence Erlbaum Associates.

Reiss, M., Behr, M., Lesh, R., & Post, T. (1985). Cognitive processes and products in proportional reasoning. In L. Streefland (Ed.), *Proceedings of the 9th Conference of the International Group for the Psychology of Mathematics Education* (Vol. 1, pp. 342–351). Noordwijkerhout, The Netherlands: Utrecht University.

Reusser, K. (1988). Problem solving beyond the logic of things: Contextual effects on understanding and solving word problems. *Instructional Science, 17*, 309–338.

Reusser, K., & Stebler, R. (1997a). Every word problem has a solution: The suspension of reality and sense-making in the culture of school mathematics. *Learning and Instruction, 7*, 309–328.

Reusser, K., & Stebler, R. (1997b). Realistic mathematical modelling through the solving of performance tasks. *Paper presented at the 7th European Conference on Learning and Instruction*, Athens, Greece.

Rogalski, J. (1982). Acquisition de notions relatives à la dimensionalité des mesures spatiales (longueur, surface) [Acquisition of notions relative to the dimensionality of spatial measures (length, area)]. *Recherches en Didactique des Mathématiques, 3*(3), 343–396.

Romberg, T. A., & Kaput, J. J. (1999). Mathematics worth teaching, mathematics worth understanding. In E. Fennema & T. A. Romberg (Eds.), "Mathematics classrooms that promote understanding" (pp. 3–19). Mahwah, NJ: Lawrence Erlbaum Associates.

Rouche, N. (1989). Prouver: amener à l'évidence ou contrôler des implications? [Proving: bringing back to evidence or verifying the implications?]. In Commission inter-IREM Histoire et Epistémologie des Mathématiques (Ed.), *La démonstration dans l'histoire* [Demonstration in history] (pp. 8–38). Lyon: IREM.

Rouche, N. (1992a). [Review of the book *Why math?*]. *Bulletin de la Société de Mathématique de Belgique (Série A), 44*(2), 245–246.

Rouche, N. (1992b). *Le sens de la mesure* [The sense of measurement]. Bruxelles: Didier Hatier.

Rouche, N. (2001, November). *La linéarité dans l'enseignement des mathématiques* [Linearity in the teaching of mathematics]. Exposé et discussion, Mons, Belgium.

Salomon, G. (1981). *Communication and education, social and psychological interactions*. Beverly Hills, CA: Sage Publications.

Sastre, M. T. M., & Mullet, E. (1998). Evolution of the intuitive mastery of the relationship between base, exponent and number magnitude in high school students. *Mathematical Cognition, 4*(1), 67–77.

Schoenfeld, A. H. (1988). When good teaching leads to bad results: The disasters of 'well-taught' mathematics courses. *Educational Psychologist, 23*, 145–166.

Schoenfeld, A. H. (1992). Learning to think mathematically: Problem solving, metacognition, and sense making in mathematics. In D. A. Grouws (Ed.), *Handbook of research on mathematics teaching and learning* (pp. 334–370). New York: Macmillan.

Schwartz, D. L., & Moore, J. L. (1998). On the role of mathematics in explaining the material world: Mental models for proportional reasoning. *Cognitive Science, 22*, 471–516.

Shaughnessy, J. M. (1992). Research in probability and statistics: Reflections and directions. In D. A. Grouws (Ed.), *Handbook of research in mathematics teaching and learning* (pp. 465–494). New York: Macmillan.

Siegler, R. S., & Jenkins, E. A. (1989). *How do children discover new strategies?* Mahwah, NJ: Lawrence Erlbaum Associates

Siegler, R. S., & Opfer, J. E. (2003). The development of numerical estimation: Evidence for multiple representations of numerical quantity. *Psychological Science, 14*, 237–243.

Silver, E. A. (1994). On mathematical problem posing. *For the Learning of Mathematics, 14,* 19–28.

Silver, E. A., & Cai, J. (1996). An analysis of arithmetic problem posing by middle school students. *Journal for Research in Mathematics Education, 27,* 521–539.

Silver, E. A., Shapiro, L. J., & Deutsch, A. (1993). Sense making and the solution of division problems involving remainders: An examination of middle school students' solution processes and their interpretations of solutions. *Journal for Research in Mathematics Education, 24,* 117–135.

Simon, M. A., & Blume, G. W. (1994). Building and understanding multiplicative relationships: A study of prospective elementary teachers. *Journal for Research in Mathematics Education, 25*, 472–494.

Sloman, S. A. (1996). The empirical case for two systems of reasoning. *Psychological Bulletin, 119*, 3–22.

Smith, D. E. (1923). *History of mathematics: General survey of the history of elementary mathematics*. New York: Dover Publications.

Spinillo, A. G., & Bryant, P. E. (1999). Proportional reasoning in young children : Part-part comparisons about continuous and discontinuous quantity. *Mathematical Cognition, 5*, 181–197.

Stacey, K. (1989). Finding and using patterns in linear generalising problems. *Educational Studies in Mathematics, 20*, 147–164.

Stanovich, K. E., & West, R. F. (2000). Individual differences in reasoning: Implications for the rationality debate. *Behavioral and Brain Sciences, 23*, 645–729.

Streefland, L. (1984). Search for the roots of ratio: Some thoughts on the long term learning process (Towards … a theory). Part I: Reflections on a teaching experiment. *Educational Studies in Mathematics, 15*, 327–348.

Székely, G. J. (1986). *Paradoxes in probability theory and mathematical statistics*. Dordrecht: Reidel.

Tate, W. E. (1994). Race, retrenchment, and the reform of school mathematics. *Phi Delta Kappan, 75*, 477–485.

Thompson, C. S., & Bush, W. S. (2003). Improving middle school teachers' reasoning about proportional reasoning. *Mathematics Teaching in the Middle School, 8*(8), 398–403.

Thompson, D. A. (1961). *On growth and form*. New York: Cambridge University Press.

Thorndike, E., Cobb, M., Orleans, J., Symonds, P., Wald, E., & Woodyard, E. (1924). *The psychology of algebra*. New York: Macmillan.

Tierney, C., Boyd, C., & Davis, G. (1990). Prospective primary teachers' conceptions of area. In G. Booker, P. Cobb, & T. N. de Mendicuti (Eds.), *Proceedings of the 14th Conference of the International Group for the Psychology of Mathematics Education* (Vol. 2, pp. 307–314). Oaxtepex, Mexico.

Tirosh, D., & Stavy, R. (1999a). Intuitive rules: A way to explain and predict students' reasoning. *Educational Studies in Mathematics*, *38*, 51–66.

Tirosh, D., & Stavy, R. (1999b). Intuitive rules and comparison tasks. *Mathematical Thinking and Learning*, *1*(3), 179–194.

Tourniaire, F., & Pulos, S. (1985). Proportional reasoning : A review of the literature. *Educational Studies in Mathematics, 16*, 181–204.

Treffers, A. (1987). *Three dimensions. A model of goal and theory description in mathematics instruction. The Wiskobas project*. Dordrecht: Reidel.

Truran, J. M. (1994). Examination of a relationship between children's estimation of probabilities and their understanding of proportion. In J. da Ponte & J. F. Matos (Eds.), *Proceedings of the 18th Conference of the International Group for the Psychology of Mathematics Education* (Vol. 4, pp. 337–344), Lisbon, Portugal.

Usiskin, Z. (2004). The arithmetic operations as mathematical models. In H.-W. Henn & W. Blum (Eds.), *ICMI Study 14: Applications and modelling in mathematics education – Pre-conference volume* (pp. 279–284). Dortmund, Germany.

Usiskin, Z. (2006). The artithmetic operations as mathematical models. In W. Blum, P. Galbraith, H.-W. Henn, & M. Niss (Eds.), *New ICMI Studies Series no.10: Applications and modelling in Mathematics Education*. New York: Springer.

Van Deyck, B. (2001). Correlatie en regressie: Een lesmodule voor de behandeling van een statistisch probleem in de derde graad van het secundair onderwijs [Correlation and regression: A lesson module for treating a statistical problem in the tertiary grades of secondary education]. *Unpublished master's thesis, University of Leuven, Belgium.*

Van Dooren, W., De Bock, D., De Bolle, E., Janssens, D., & Verschaffel, L. (2003a). The illusion of linearity: The role of direct versus indirect perimeter and area measures. *Mediterranean Journal for Research in Mathematics Education, 2*, 1–18.

Van Dooren, W., De Bock, D., Depaepe, F., Janssens, D., & Verschaffel, L. (2003b). The illusion of linearity: Expanding the evidence towards probabilistic reasoning. *Educational Studies in Mathematics, 53*, 113–138.

Van Dooren, W., De Bock, D., Hessels, A., Janssens, D., & Verschaffel, L. (2005). Not everything is proportional: Effects of age and problem type on propensities for overgeneralization. *Cognition and Instruction, 23*(1), 57–86.

Van Dooren, W., De Bock, D., Hessels, A., Janssens, D., & Verschaffel, L. (2004). Remedying secondary school students' illusion of linearity: A teaching experiment aiming at conceptual change. *Learning and Instruction, 14*, 485–501.

Van Dooren, W., De Bock, D., Weyers, D., & Verschaffel, L. (2004). Challenging the predictive power of intuitive rules: A replication and extension study on the impact of 'More A–more B' and 'Same A–same B'. *Educational Studies in Mathematics, 56*, 179–207.

Van Essen, G., & Hamaker, C. (1990). Using self-generated drawings to solve arithmetic word problems. *Journal of Educational Research, 83*(6), 301–312.

Vergnaud, G. (1983). Multiplicative structures. In R. Lesh & M. Landau (Eds.), *Acquisition of mathematics concepts and processes* (pp. 127–174). New York: Academic Press.

Verschaffel, L. (2002). Taking the modelling perspective seriously at the elementary school level: Promises and pitfalls. In A. D. Cockburn & E. Nardi (Eds.), *Proceedings of the 26th Conference of the International Group for the Psychology of Mathematics Education* (Vol. 1, pp. 64–80). Norwich, U.K.

Verschaffel, L., & De Corte, E. (1997). Teaching realistic modelling in the elementary school. A teaching experiment with fifth graders. *Journal for Research in Mathematics Education, 28*, 577–601.

Verschaffel, L., De Corte, E., & Lasure, S. (1994). Realistic considerations in mathematical modelling of school arithmetic word problems. *Learning and Instruction, 4*, 273–294.

Verschaffel, L., De Corte, E., Lasure, S., Van Vaerenbergh, G., Bogaerts, H., & Ratinckx, E. (1999). Design and evaluation of a learning environment for mathematical modelling and problem solving in upper elementary school children. *Mathematical Thinking and Learning, 1*, 195–229.

Verschaffel, L., De Corte, E., Lowyck J., Dhert S., & Vandeput, L. (2000). Supporting mathematical problem solving and posing in upper elementary school children by means of Knowledge Forum. (Deliverable of project No. 2017 CL-Net: *Computer Supported Collaborative Learning Networks in Primary and Secondary Education.*) Leuven, Belgium: Center for Instructional Psychology and Technology, University of Leuven.

Verschaffel, L., Greer, B., & De Corte, E. (2000). *Making sense of word problems*. Lisse, The Netherlands: Swets & Zeitlinger.

Verschaffel, L., Greer, B., & De Corte, E. (2002). Everyday knowledge and mathematical modeling of school word problems. In K. Gravemeijer, R. Lehrer, B. van Oers, & L. Verschaffel (Eds.), *Symbolizing, modeling and tool use in mathematics education* (pp. 257–276). Dordrecht: Kluwer.

Viennot, L. (1979). *Le raisonnement spontané en dynamique élémentaire* [Spontaneous reasoning in elementary dynamics]. Paris: Hermann.

Villarreal, M., Esteley, C., & Alagia, H. (2004). University students' extension of linear models to non-linear situations. In M. J. Høines & A. B. Fuglestad (Eds.), *Proceedings of the 28th Conference of the International Group for the Psychology of Mathematics Education* (Vol. 1, p. 364). Bergen, Norway.

Vlahovic-Stetic, V. (1999). Word-problem solving as a function of problem type, situational context and drawing. *Studia Psychologica, 41*(1), 49–62.

Vosniadou, S. (1994). Knowledge representation and organization. In T. Husen & N. Postlethwaite (Eds.), *The international encyclopedia of education* (2nd edition, pp. 3 151–3 155). Oxford: Pergamon.

Vosniadou, S. (1999). Conceptual change research: State of the art and future directions. In W. Schnotz, S. Vosniadou, & M. Carretero (Eds.), *New perspectives on conceptual change* (pp. 3–13). Oxford: Pergamon.

Vosniadou, S., Ioannides, C., Dimitrakopoulou, A., & Papademetriou, E. (2001). Designing learning environments to promote conceptual change in science. *Learning and Instruction, 11*, 381–419.

Weller, H. G. (1995). Diagnosing and altering three Aristotelian alternative conceptions in dynamics: Microcomputer simulations of scientific models. *Journal of Research in Science Teaching, 32*, 271–290.

Wertheimer, M. (1945). *Productive thinking*. New York: Harper & Brothers.

Wyndhamn, J., & Säljö, R. (1997). Word problems and mathematical reasoning: A study of children's mastery of reference and meaning in textual realities. *Learning and Instruction, 7*(4), 361–382.

Yackel, E., & Cobb, P. (1996). Sociomathematical norms, argumentation, and autonomy in mathematics. *Journal for Research in Mathematics Education, 27*, 458–477.

Index

Mathematics Education Library

Managing Editor: A.J. Bishop, Melbourne, Australia

1. H. Freudenthal: *Didactical Phenomenology of Mathematical Structures.* 1983
   ISBN 90-277-1535-1 HB; 90-277-2261-7 PB
2. B. Christiansen, A. G. Howson and M. Otte (eds.): Perspectives on Mathematics Education. Papers submitted by Members of the Bacomet Group. 1986
   ISBN 90-277-1929-2 HB; 90-277-2118-1 PB
3. A. Treffers: Three Dimensions. A Model of Goal and Theory Description in Mathematics Instruction TheWiskobas Project. 1987 ISBN 90-277-2165-3
4. S. Mellin-Olsen: The Politics of Mathematics Education. 1987
   ISBN 90-277-2350-8
5. E. Fischbein: Intuition in Science and Mathematics. An Educational Approach. 1987 ISBN 90-277-2506-3
6. A.J. Bishop: Mathematical Enculturation. A Cultural Perspective on Mathematics Education.1988 ISBN 90-277-2646-9 HB; 1991 0-7923-1270-8 PB
7. E. von Glasersfeld (ed.): Radical Constructivism in Mathematics Education. 1991 ISBN 0-7923-1257-0
8. L. Streefland: Fractions in Realistic Mathematics Education. A Paradigm of Developmental Research. 1991 ISBN 0-7923-1282-1
9. H. Freudenthal: Revisiting Mathematics Education. China Lectures. 1991
   ISBN 0-7923-1299-6
10. A.J. Bishop, S. Mellin-Olsen and J. van Dormolen (eds.): Mathematical Knowledge: Its Growth Through Teaching. 1991 ISBN 0-7923-1344-5
11. D. Tall (ed.): Advanced Mathematical Thinking. 1991 ISBN 0-7923-1456-5
12. R. Kapadia and M. Borovcnik (eds.): Chance Encounters: Probability in Education. 1991 ISBN 0-7923-1474-3
13. R. Biehler, R.W. Scholz, R. Sträßer and B. Winkelmann (eds.): Didactics of Mathematics as a Scientific Discipline. 1994 ISBN 0-7923-2613-X
14. S. Lerman (ed.): Cultural Perspectives on the Mathematics Classroom. 1994
    ISBN 0-7923-2931-7
15. O. Skovsmose: Towards a Philosophy of Critical Mathematics Education. 1994
    ISBN 0-7923-2932-5
16. H. Mansfield, N.A. Pateman and N. Bednarz (eds.): Mathematics for Tomorrow's Young Children. International Perspectives on Curriculum. 1996
    ISBN 0-7923-3998-3
17. R. Noss and C. Hoyles:Windows on Mathematical Meanings. Learning Cultures and Computers. 1996 ISBN 0-7923-4073-6 HB; 0-7923-4074-4 PB
18. N. Bednarz, C. Kieran and L. Lee (eds.): Approaches to Algebra. Perspectives for Research and Teaching.1996 ISBN 0-7923-4145-7 HB; 0-7923-4168-6, PB
19. G. Brousseau: Theory of Didactical Situations in Mathematics. Didactique des Mathématiques 1970-1990. Edited and translated by N. Balacheff, M. Cooper, R. Sutherland and V. Warfield. 1997 ISBN 0-7923-4526-6
20. T. Brown: Mathematics Education and Language. Interpreting Hermeneutics and Post-Structuralism. 1997 ISBN 0-7923-4554-1 HB
    Second Revised Edition. 2001 ISBN 0-7923-6969-6 PB

Mathematics Education Library

21. D. Coben, J. O'Donoghue and G.E. FitzSimons (eds.): Perspectives on Adults Learning Mathematics. Research and Practice. 2000 ISBN 0-7923-6415-5
22. R. Sutherland, T. Rojano, A. Bell and R. Lins (eds.): Perspectives on School Algebra. 2000 ISBN 0-7923-6462-7
23. J.-L. Dorier (ed.): On the Teaching of Linear Algebra. 2000 ISBN 0-7923-6539-9
24. A. Bessot and J. Ridgway (eds.): Education for Mathematics in the Workplace. 2000 ISBN 0-7923-6663-8
25. D. Clarke (ed.): Perspectives on Practice and Meaning in Mathematics and Science Classrooms. 2001 ISBN 0-7923-6938-6 HB; 0-7923-6939-4 PB
26. J. Adler: Teaching Mathematics in Multilingual Classrooms. 2001 ISBN 0-7923-7079-1 HB; 0-7923-7080-5 PB
27. G. de Abreu, A.J. Bishop and N.C. Presmeg (eds.): Transitions Between Contexts of Mathematical Practices. 2001 ISBN 0-7923-7185-2
28. G.E. FitzSimons:What Counts as Mathematics? Technologies of Power in Adult and Vocational Education. 2002 ISBN 1-4020-0668-3
29. H.Alrø andO. Skovsmose:Dialogue and Learning in Mathematics Education. Intention, Reflection,Critique.2002 ISBN 1-4020-0998-4 HB;1-4020-1927-0 PB
30. K. Gravemeijer, R. Lehrer, B. van Oers and L. Verschaffel (eds.): Symbolizing, Modeling and Tool Use in Mathematics Education. 2002 ISBN 1-4020-1032-X
31. G.C. Leder, E. Pehkonen and G. T rner (eds.): Beliefs: A Hidden Variable in Mathematics Education? 2002 ISBN 1-4020-1057-5 HB; 1-4020-1058-3 PB
32. R. Vithal: In Search of a Pedagogy of Conflict and Dialogue for Mathematics Education. 2003 ISBN 1-4020-1504-6
33. H.W. Heymann: Why Teach Mathematics? A Focus on General Education. 2003 Translated by Thomas LaPresti ISBN 1-4020-1786-3
34. L. Burton: Mathematicians as Enquirers: Learning about Learning Mathematics. 2004 ISBN 1-4020-7853-6 HB; 1-4020-7859-5 PB
35. P. Valero, R. Zevenbergen (eds.): Researching the Socio-Political Dimensions of Mathematics Education: Issues of Power in Theory and Methodology. 2004 ISBN 1-4020-7906-0
36. D. Guin, K. Ruthven, L. Trouche (eds.) The Didactical Challenge of Symbolic Calculators: Turning a Computational Device into a Mathematical Instrument 2005 ISBN 0-387-23158-7
37. J. Kilpatrick, C. Hoyles, O. Skovsmose (eds. in collaboration with Paola Valero) Meaning in Mathematics Education. 2005 ISBN 0-387-24039-X
38. H. Steinbring: The Construction of New Mathematical Knowledge in Classroom Interaction: An Epistemological Perspective. 2005. ISBN 0-387-24251-1
39. M.Borba, M. Villarreal: Humans-with-Media and the Reorganization of Mathematical Thinking: Information and Communication Technologies, Modeling, Visualization and Experimentation. 2005 ISBN 0-387-24263-5 HB, ISBN 0-387-32821-1 PB
40. G. Jones (ed): Exploring Probability in School: Challenges for Teaching and Learning. 2005 ISBN 0-387-24529-4
41. D. DeBock, W. Van Dooren, D. Janssens, and L. Verschaffel: The Illusion of Linearity: From Analysis to Improvement. 2007 ISBN 978-0-387-71082-2

Printed in the United States
100491LV00001B/46/A

9 780387 710822